THE FORM OF FINITE GROUPS:

A Course on Finite Group Theory

The Form of Finite Groups:
A Course on Finite Group Theory

Stephen G. Odaibo
M.S.(Math), M.S.(Comp. Sci.), M.D.

Quantum Lucid Research Laboratories
Dubuque, IA

First published in the United States of America by
Symmetry Seed Books.

For information about permissions to reproduce selections
from this book, send an email to: finitegroups@qlucid.com

Library of Congress Control Number: 2015920384
Symmetry Seed Books, Dubuque, IA.

ISBN-13: 978-0997116304

ISBN-10: 0997116307

Acknowledgements and Dedication

Many moons ago I was in the Math Fast-Track Program at the University of Alabama–Birmingham. I had outstanding mentors to whom I am grateful. My love of and commitment to Mathematics was cultivated during my time in the Program and has endured the years. This explains in part how I have done here what some may deem an improbable task for a physician — writing a book on finite group theory. I especially thank Dr. James Robert Ward, Jr., Dr. Marius Nkashama, Dr. John Mayer, Dr. Lex Oversteegen, and all the faculty in the UAB Math department. Dr. James Robert Ward, Jr. has since left this world. He died on March 25th 2013. He was a very kind-hearted professor who took great interest in his students' well-being. His area was differential equations and he mentored me on an enzyme kinetics modeling project. As an undergrad, I would often stop-by his office just to chat. He wrote me an excellent letter of recommendation for medical school, and remained interested in my progress afterwards. He was good-natured and of good humor. He was a professor, a mentor, and a friend. He is dearly missed. All proceeds of this book will go to the Dr. James Robert Ward, Jr. Memorial Scholarship Fund which was set-up by his lovely wife Ana.

Feedback and Errata

Please kindly email suggestions, errors, and other feedback
to finitegroups@qlucid.com

Contents

Preface

This book's target audience is broad. The book should be a great resource for any lover of knowledge curious about the underpinnings of our physical world, and with a serious desire to understand it. For this person, a familiarity with Linear algebra, basic set theory, and some exposure to proof writing are recommended prerequisites. Others for whom this book is intended include: (i) undergraduate students who have completed a course in Linear algebra, basic set theory, and have some exposure to proof writing, (ii) Mathematics graduate students preparing for the Group theory part of their Algebra qualifying exams, and (iii) researchers in Physical science disciplines such as Physics, Chemistry, Computer science, Statistics, or Engineering whose work require Group theory.

The exercises at the end of each chapter are deliberately accessible. After reading through the chapter, one should be able to tackle the problems at the end, and in a good number of cases the solution can be found in the text itself. This is intentionally done to prevent the reader from getting unduly stuck and frustrated. It also allows the reader to develop a quick understanding and rapidly verify that understanding.

Several of the chapters open with an interlude of poetry, a quote, or a biography of a mathematician who contributed to the development or discovery of the chapter's contents. Diagrammatic illustrations are presented throughout the book to help illuminate the concepts.

This book is not an exhaustive reference of Finite group

theory. Its 225 pages cover the key concepts of Group theory, and should leave one very well-prepared to go further in Algebra into the study of Rings, Fields, Modules, and Galois theory. Needless to say, some topics were left out of the book. For instance, the chapter on the applications of the Sylow theorems covers only the most important cases. It does give one a full taste and equipment for how the Sylow theorems can be applied to elucidate group structure. A full exposition of the consequences of the Sylow theorems would include the full classification of finite simple groups, and could easily fill a 4,000 page volume by itself. This was certainly not the intention of this book.

What the reader will find in this book is a clear, accessible, and well-written exposition of finite group theory. It covers the core topics such a quotient groups and Lagrange's theorem, the symmetric group, Cayley's theorem, Cauchy's theorem, the Sylow theorems, the Fundamental theorem of finite abelian groups, and an overview of groups of small order. This book will provide the reader with the skills and requisite knowledge for research problem solving and for further study in Algebra. And it will provide the lover of knowledge and of nature with a fluency in the language in which nature is expressed.

—Stephen G. Odaibo, M.D.

Chapter 1

Introduction

1.1   Why Understand Group Theory?

So why understand group theory? Like good poetry, art, or music, it is beautiful and it fits. However there are many reasons to study group theory other than for its astonishing beauty and elegance. Nonetheless, its beauty and elegance are reason enough and arguably underly the many other conceivable reasons. Albert Einstein once said "I have deep faith that the principle of the universe will be beautiful and simple". Similarly, the physics Nobel laureate Murray Gell-Mann remarked that beauty and elegance are design principles of nature. Our perception of beauty is often intrinsically linked to the notion of symmetry, and symmetry in some sense is what group theory is all about. Look around you, at a sunrise or sunset. Look at the perfect symmetry of a rose petal. Look at the oval-shaped dew drop that glistens like a blue topaz atop the rose's leaves. Look at a rainbow or at the fading streaks of the Aurora Borealis. It is undeniable that the concepts of symmetry, number structure, and constraints of finiteness come into play fundamentally in every aspect of our physical world.

It is conceivable that group theory is the language in which our physical world was spoken into being. This is clearly apparent in modern physics. Group theory is the language of electrons, protons, quarks, and other subatomic particles. It is the language of crystal structure in chemistry. It is the language of Einstein's theory of relativity, whose dialect is the Lorenz group of transformations. The list goes on and on, and is complete. To understand nature, one must understand group theory.

Examples of groups include the real numbers, the integers, invertible matrices, and flips of a switch. This could be a light switch on your bathroom wall, or a molecular switch in your genetic code which either starts or stops a disease process.

Yes, group theory is important. But why understand *finite* group theory? The answer is simple. We live in a finite world. At any given time, the total number of molecules of water in all the earth's oceans, seas, rivers, lakes, and atmosphere is finite. It is a single number! The number of planets in our galaxy is finite. Our planet has one sun, one moon, and two poles. There are three vertices in a triangle, five petals in an asteracea, and 23 pairs of chromosomes in humans. Somewhat ironically, this constraint of finiteness confers greater structure and complexity on the subject. A boomerang thrown in a small confined space has a much more complicated trajectory than one thrown in the open air. Yes, nature is beautiful, but can be complex. The structure, relationships, and consequences of finite group theory are well-suited to model the complexity in many a natural phenomenon. We will see in the later chapters how the number of elements in a group determine its structure and its form. We will see how for instance a group of prime order must be cyclic like a bicycle wheel. And we will study the Sylow theorems, which based only on the number of elements in a group, tell much about the number and types of subgroups in the group.

The number systems are the blueprint of finite groups. The real number system, the integers within them, the distribution of prime numbers, divisibility, factorization and so on, are the instructions and constraints on the structure of groups. As a direct consequence, attributes of the number systems also encode the natural phenomena which these groups model and describe. In some sense the number systems are the "DNA" of group theory and hence they are the "DNA" of nature itself. Carl Friedrich Gauss commented that "mathematics is the queen of the sciences and number theory is the queen of mathematics". The *finite* adjective in the title and content of this book is *a* gate into this world of numbers.

Indeed every branch of science seeks to understand the structure and relationships of some aspect of our physical world. And fluency in finite group theory is requisite for the task. To understand nature, one must understand *finite* group theory.

1.2   What is a Group?

Groups are characterized by a number of fundamental properties and relationships.[1-5] To understand group theory is to
develop a palpable sense of these properties and relationships.
Here, we start by defining what it means to be a group. And
we examined some examples of groups. So without further
ado, what is a group?

Definition 1.1. A group G is a set of algebraic objects which
is equipped with a binary operation called the *group multiplication*, and which has the following four properties:

1. Identity: There exists an identity element, e, such that
   $g \cdot e = e \cdot g = g$ for all $g \in G$.

2. Closure under multiplication: The group is closed under
   multiplication. This means that for any $g, h \in G$ it
   holds true that $g \cdot h \in G$.

3. Associativity: The group multiplication is associative.
   This means that for any $g, h, k \in G$, it holds true that
   $g \cdot (h \cdot k) = (g \cdot h) \cdot k$.

4. Inverse: For any element $g \in G$ there exists an inverse
   $g^{-1} \in G$ such that $gg^{-1} = g^{-1}g = e$.

   The group G is sometimes denoted $(G, \cdot)$ to highlight the
group multiplication. Note that the term "group multiplication" is an abstract term which does not necessarily mean
"times". It can be any binary operation consistent with
the other requirements of group. Examples include "plus"
or "convolution" or "matrix matrix multiplication". For in-
stance $(\mathbb{R}, +)$ denotes the reals under addition, while $(\mathbb{R}, \times)$
denotes the reals under multiplication. Note also that there
is great diversity of notational representation in the litera-
ture and in textbooks. For instance the above examples are
often denoted $\mathbb{R}^+$ and $\mathbb{R}^\times$ respectively. The binary operation

is often omitted as we will also often do in this book. For instance, instead of $g \cdot h$ we will often just write gh.

1.3 Examples of Groups

Example 1.2. $(\mathbb{R}, +)$ is a group.

Proof: To see this, we just have to verify that it satisfies each of the above four properties. The identity here is 0. See that $0 + r = r + 0 = r$ for all $r \in \mathbb{R}$. To verify closure under multiplication note that for any $r, s \in \mathbb{R}$, it is true that $r + s \in \mathbb{R}$. To verify associativity, note that $(r + s) + t = r + (s + t)$ for any $r, s, t \in \mathbb{R}$. To verify the existence of an inverse, note that $r + (-r) = (-r) + r = 0$ for any $r \in \mathbb{R}$.

Example 1.3. $(\mathbb{R}, \times)$ is a group.

Proof: Here, the identity is 1. See that $r \times 1 = 1 \times r = r$ for any $r \in \mathbb{R}$. To verify closure under multiplication note that $r \times s \in \mathbb{R}$ for any $r, s \in \mathbb{R}$. To verify associativity, note that $(r \times s) \times t = r \times (s \times t)$ for any $r, s, t \in \mathbb{R}$. To verify the existence of an inverse, note that $r \times (1/r) = (1/r) \times r = 1$ for any $r \in \mathbb{R}$.

Example 1.4. $(\mathbb{Z}, +)$, i.e. the integers under addition — is a group.

Proof: We know that $\mathbb{Z} \subset \mathbb{R}$, therefore since we have shown that $(\mathbb{R}, +)$ is a group all we have left to show is closure under multiplication. This follows trivially since the sum of any two integers is also an integer.

Example 1.5. $(\mathbb{Z}, \times)$, that is, the integers under multiplication — is a group.

Proof: Same as for $(\mathbb{Z}, +)$ above.

Example 1.6. The set of invertible 2×2 matrices with real entries is a group. We denote this set $GL_2(\mathbb{R})$. The group operation is matrix matrix multiplication.

Proof: The identity is given by the 2×2 identity matrix,

$$\begin{pmatrix} 1 & 0 \\ 0 & 1 \end{pmatrix}. \tag{1.1}$$

We can easily check that for any matrix $\begin{pmatrix} a & b \\ c & d \end{pmatrix} \in GL_2(\mathbb{R})$ it holds that,

$$\begin{pmatrix} 1 & 0 \\ 0 & 1 \end{pmatrix} \begin{pmatrix} a & b \\ c & d \end{pmatrix} = \tag{1.2}$$

$$\begin{pmatrix} a & b \\ c & d \end{pmatrix} \begin{pmatrix} 1 & 0 \\ 0 & 1 \end{pmatrix} =$$

$$\begin{pmatrix} a & b \\ c & d \end{pmatrix}.$$

To verify associativity, observe the following outcome of multiplying three arbitrary elements of $GL_2(\mathbb{R})$..

$$\left[\begin{pmatrix} a & b \\ c & d \end{pmatrix} \begin{pmatrix} e & f \\ g & h \end{pmatrix} \right] \begin{pmatrix} i & j \\ k & l \end{pmatrix} = (1.3)$$

$$\begin{pmatrix} ae+bg & af+bh \\ ce+dg & cf+dh \end{pmatrix} \begin{pmatrix} i & j \\ k & l \end{pmatrix} =$$

$$\begin{pmatrix} (ae+bg)i+(af+bh)k & (ae+bg)j+(af+bh)l \\ (ce+dg)i+(cf+dh)k & (ce+dg)j+(cf+dh)l \end{pmatrix} =$$

$$\begin{pmatrix} aei+bgi+afk+bhk & aej+bgj+afl+bhl \\ cei+dgi+cfk+dhk & cej+dgj+cfl+dhl \end{pmatrix} =$$

$$\begin{pmatrix} a(ei+fk)+b(gi+hk) & a(ej+fl)+b(gj+hl) \\ c(ei+fk)+d(gi+hk) & c(ej+fl)+d(gj+hl) \end{pmatrix} =$$

$$\begin{pmatrix} a & b \\ c & d \end{pmatrix} \begin{pmatrix} ei+fk & ej+fl \\ gi+hk & gj+hl \end{pmatrix} =$$

$$\begin{pmatrix} a & b \\ c & d \end{pmatrix} \left[\begin{pmatrix} e & f \\ g & h \end{pmatrix} \begin{pmatrix} i & j \\ k & l \end{pmatrix} \right].$$

To verify the existence of inverses, we only have to trivially return to the definition of $GL_2(\mathbb{R})$. It is the set of *invertible* 2×2 matrices with real entries. Therefore by definition, each element of $GL_2(\mathbb{R})$ has an inverse in $GL_2(\mathbb{R})$.

To verify closure under multiplication, we note that multiplication of two 2×2 real-entried matrices yields a 2×2 real-entried matrix. All that remains is to show that when the multiplied matrices are invertible then the product is also invertible. Consider two invertible matrices

$$\begin{pmatrix} a & b \\ c & d \end{pmatrix} \quad \text{and} \quad \begin{pmatrix} e & f \\ g & h \end{pmatrix}. \tag{1.4}$$

Their product is,

$$\begin{pmatrix} a & b \\ c & d \end{pmatrix} \begin{pmatrix} e & f \\ g & h \end{pmatrix}.$$

Denote their inverses by

$$\begin{pmatrix} a' & b' \\ c' & d' \end{pmatrix} \quad \text{and} \quad \begin{pmatrix} e' & f' \\ g' & h' \end{pmatrix} \tag{1.5}$$

respectively. We claim that the inverse of their product is given by,

$$\begin{pmatrix} e' & f' \\ g' & h' \end{pmatrix} \begin{pmatrix} a' & b' \\ c' & d' \end{pmatrix}. \tag{1.6}$$

This is easily verified by multiplication of the product and its presumed inverse to get,

$$\begin{pmatrix} a & b \\ c & d \end{pmatrix} \begin{pmatrix} e & f \\ g & h \end{pmatrix} \begin{pmatrix} e' & f' \\ g' & h' \end{pmatrix} \begin{pmatrix} a' & b' \\ c' & d' \end{pmatrix} = \qquad (1.7)$$

$$\begin{pmatrix} a & b \\ c & d \end{pmatrix} \begin{pmatrix} 1 & 0 \\ 0 & 1 \end{pmatrix} \begin{pmatrix} a' & b' \\ c' & d' \end{pmatrix} =$$

$$\begin{pmatrix} a & b \\ c & d \end{pmatrix} \begin{pmatrix} a' & b' \\ c' & d' \end{pmatrix} =$$

$$\begin{pmatrix} 1 & 0 \\ 0 & 1 \end{pmatrix}.$$

In the above equation, we used associativity to collapse the middle terms to the identity, and then simplified onwards. The above construction of the inverse generalizes. It is true that if G is a group and $g, h \in G$, then

$$(gh)^{-1} = h^{-1}g^{-1}. \qquad (1.8)$$

The group we just described belongs to a larger class of groups called the generalized linear group and denoted $GL_n(\mathbb{R})$. The "n" denotes that the group elements are invertible $n \times n$ matrices and the "$\mathbb{R}$" denotes that the matrix entries are real. Of note, an even more general category is $GL_n(\mathbb{F})$ where $\mathbb{F}$ is any specified field, for example the complex number field $\mathbb{C}$ or the field of integers $\mathbb{Z}$.

Example 1.7. Let $S^1 = \{z \in \mathbb{C} \text{ such that } |z| = 1\}$ where $|z| = (\Re[z])^2 + (\Im[z])^2$ is the length of z. $\Re[z]$ and $\Im[z]$ denote the real and imaginary parts of z respectively. S^1 is a group. It is called the unit circle group because its elements are the points on the unit circle centered about the origin in the complex plane. The group operation is complex number multiplication.

Proof: Let $z \in S^1$ and $z = a + bi$, then $|z| = a^2 + b^2 = 1$. To see that S^1 is a group, note the following. The identity is $e = 1 + 0i = 1$, and one readily sees that $z \times 1 = 1 \times z = z$. Associativity of S^1 is inherited directly from the associativity of $(\mathbb{C}, \times)$. To show closure under inverses, we claim $z^{-1} = a - bi$. First we show that $a - bi \in S^1$:

$$a^2 + (-b)^2 = a^2 + b^2 = 1. \tag{1.9}$$

Next to see that $z^{-1} = a - bi$ we evaluate the following,

$$
\begin{aligned}
(z)(a - bi) &= \\
(a + bi)(a - bi) &= \\
a^2 - abi + abi + b^2 &= \\
a^2 + b^2 &= 1.
\end{aligned}
\tag{1.10}
$$

To verify closure under multiplication, let $z_1 = (a + bi)$ and $z_2 = (c + di)$ be arbitrary elements of S^1. Note that $a^2 + b^2 = c^2 + d^2 = 1$. We need to evaluate the product $z_1 z_2$ and then check to see if its length is 1.

$$
\begin{aligned}
z_1 z_2 &= \\
(a + bi)(c + di) &= \\
ac + adi + bci - bd &= \\
(ac - bd) + (ad + bc)i.
\end{aligned}
\tag{1.11}
$$

Next we check to see if $z_1 z_2$ is of unit length.

$$
\begin{aligned}
|z_1 z_2| &= \\
(\Re[z_1 z_2])^2 + (\Im[z_1 z_2])^2 &= \\
(ac - bd)^2 + (ad + bc)^2 &= \\
(a^2 c^2 - 2acbd + b^2 d^2) + (a^2 d^2 + 2adbc + b^2 c^2) &= \\
a^2(c^2 + d^2) - 2adbc + 2adbc + b^2(c^2 + d^2) &= \\
a^2(c^2 + d^2) + b^2(c^2 + d^2) &= \\
(a^2 + b^2)(c^2 + d^2) &= \\
1 \times 1 &= \\
1.
\end{aligned}
\tag{1.12}
$$

This shows closure under multiplication and completes the proof that S^1 is a group. $\boxed{\text{QED}\checkmark}$

All the above examples are infinite groups. For example there are an infinite number of real numbers, an infinite number of integers, an infinite number of invertible 2×2 matrices, and an infinite number of points on the unit circle centered about the origin. Though this book is about finite groups, the fundamental character and definition of a group does not discriminate between the finite and the infinite. The additional structure and complexity conferred by boundedness becomes more manifest and relevant as the book proceeds. Let us now consider some examples of finite groups.

Example 1.8. The following set is a group: $\{e, a, b, c\}$ where e is the identity, $a^2 = b^2 = c^2 = e$, $ab = c$, $bc = a$, and $ac = b$.

Proof: By definition the group contains the identity, contains inverses for all its elements, and is closed under multiplication. All that remains to show explicitly is associativity. The product of any three non-trivial elements yields the identity which is unique, proving associativity. E.g.

$$(ab)c = c^2 = e = a^2 = a(bc). \tag{1.13}$$

This four element group is called the Klein-Four group. It is one of two four element groups that exist. The other is $\mathbb{Z}/4$. We will see more of both types of groups as we proceed through the book.

Example 1.9. The set of integers modulo 7 under addition is a group. It is denoted $\mathbb{Z}/7$. It consists of the numbers, $\{0, 1, 2, 3, 4, 5, 6\}$. The rules of 'multiplication' are such that if $a, b \in \mathbb{Z}/7$, then in $\mathbb{Z}/7$, $a + b \equiv \text{rem}(\frac{a+b}{7})$; where rem denotes the remainder. For instance $5 + 6 \equiv \text{rem}(\frac{5+6}{7}) = \text{rem}(11/7) = 4$. While $2 + 3 \equiv \text{rem}(5/7) = 5$. While $100 \equiv \text{rem}(100/7) = 2$. Let's show that $\mathbb{Z}/7$ is a group.

Proof: The identity is 0, and it is trivial to see for instance that $0 + 5 = 5 + 0 = 5$. Associativity is inherited directly from $(\mathbb{Z}, +)$. For any $g \in \mathbb{Z}/7$, $g^{-1} = 7 - g$. Therefore $g + (7 - g) = (7 - g) + g = 7 \equiv 0$. And finally, we already showed closure under multiplication in our multiplication scheme above, ire. for $g, h \in \mathbb{Z}/7$, $g + h \equiv \operatorname{rem}(\frac{g+h}{7}) \in \mathbb{Z}/7$.

The group $\mathbb{Z}/7$ belongs to the larger class of groups $\mathbb{Z}/n$ where n is any natural number. These groups are called *cyclic groups of order n*, or are called the *integers modulo n*.

Example 1.10. The set of possible permutations of four letters is a group. The group operation is composition of permutations. This group is denoted S_4.

Proof: Consider the following ordered array of letters:

$$[a_1, a_2, a_3, a_4]$$

The identity element of S_4 is the permutation that leaves the ordering of the letters unperturbed, that is e is the permutation that yields $[a_1, a_2, a_3, a_4]$. Given any permutation σ, its inverse denoted σ^{-1} is the permutation that reverses the effect of σ. To verify closure under multiplication, note that composition of any two permutations is itself a permutation. We leave the proof of associativity as an exercise. Note that this group contains $4! = 24$ elements, as there are $4!$ possible permutations of the 4 letters.

The group of permutations of n letters is called the symmetric group on n letters. It is denoted S_n and has $n!$ elements. It plays a central role in finite group theory. As we will see later in this book, every finite group of order n is a subgroup of S_n. This theorem is called Cayley's theorem.

Example 1.11. The symmetries of a square form a group. *The symmetries of a square is the answer to the question: What changes can one make to an unlabeled square so that the square appears unchanged?* Answer: we can rotate the

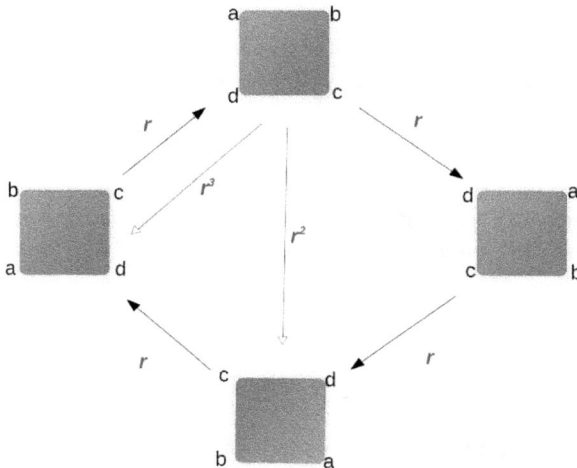

Figure 1.1: Rotational Symmetries of the Square. r denotes clockwise rotation by 90 degrees.

square by 90 degrees, 180 degrees, 270 degrees, or 360 degrees. The rotation by 360 degrees places the square back exactly where we started — this means that following a 360 degree rotation, even a labeled square appears unchanged. Therefore rotation by 360 degrees corresponds to the identity. In addition to the listed rotations, we can also reflect the square about any mid-line connecting any two edges or any two vertices. For instance we can reflect it about the hori-

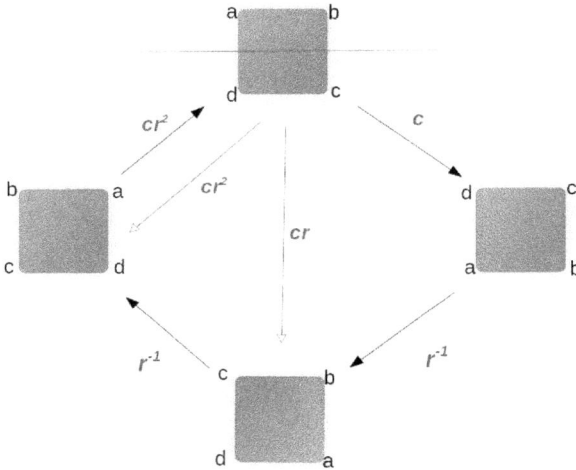

Figure 1.2: Reflection Symmetries of the Square. The red horizontal line represents the axis of reflection. Reflection about the horizontal is denoted by c, while clockwise rotation by 90 degrees is denoted by r. Multiplication is composition and therefore read right-to-left. For example cr means rotate clockwise by 90 degrees then reflect about the horizontal.

zontal, about the vertical, about the north-east to south-west diagonal, or about the north-west to south-east diagonal. Altogether, these sum up to 8 possible transformations — four rotations and four reflections. These are called symmetries because they are transformations about a line of symmetry. The square is symmetric about particular axes of rotation or reflection, and rotation or reflection about those respective axes leave the square 'unchanged'. More accurately, the unlabeled square is perceived as unchanged by an observer who is blindfolded during the transformation. Figures (1.1) and (1.2) illustrate the elements of this group. The group is called the Dihedral group of order 8. As mentioned, it contains 4 rotations (including the identity) and 4 reflections, for a total of 8 elements. And the group operation is composition. In this book we will denote this group D_8, but note that in some other texts it is denoted D_4. This group belongs to a larger class of groups called the Dihedral group of order $2n$, which represents the symmetries of the n-sided regular polygon. We will learn more about this group in Chapter (6).

1.4 Exercises

1. Are the natural numbers under addition a group? Why or why not?

2. Are the natural numbers under multiplication a group? Why or why not?

3. Given a group, $(G, \circ)$ with function composition as group operation, show that for any element of the group, left inverse must equal right inverse.

4. Show that the identity is unique.

5. Let G be a group and $x, y \in G$. What is $(xy)^{-1}$?

6. Show that the behavior of any basic on-off light switch is modeled by a group? What is this group called?

7. Describe a group that consists of only two elements. Are there any other two element groups in existence other than the group you described? If there are any other such groups, describe them.

8. Describe a group that consists of only three elements. Are there any other three element groups in existence other than the group you described? If there are any other such groups, describe them.

9. Describe a group that consists of only four elements. Are there any other four element groups in existence other than the group you described? If there are any other such groups, describe them.

10. Show that $\mathbb{Z}/9$ under addition is a group.

11. Describe any proper subsets of the elements in $\mathbb{Z}/9$ which themselves form a group. How many such proper subsets are there?

12. Show that $(\mathbb{C}, \times)$ the complex numbers under complex number multiplication is a group.

13. Show that S^1 the set of complex numbers on the unit circle centered about origin is a group, where the group operation is complex number multiplication.

14. Consider the set of permutations of 5 letters under composition. Call this set S_5.

    a) Show that S_5 is a group.

    b) How many elements are in S_5?

15. Let $GL_n(\mathbb{R})$ denote the set of invertible $n \times n$ matrices with real entries. Take matrix matrix multiplication as the binary operation between elements of this set. Show that $GL_n(\mathbb{R})$ is indeed a group.

16. Let $O_n(\mathbb{R}) = \{M \in GL_n(\mathbb{R}) \text{ such that } M^T M = I\}$, where I is the $n \times n$ identity matrix and M^T is the transpose of M. Show that $O_n(\mathbb{R})$ is a group. This group is called the orthogonal group.

17. Let $SO_n(\mathbb{R}) = \{M \in O_n(\mathbb{R}) \text{ such that } det(M) = 1\}$. Show that $SO_n(\mathbb{R})$ is a group. This group is called the special orthogonal group.

18. Show that D_6, the symmetries of the triangle, is a group.

19. Show that D_8, the symmetries of the square, is a group.

20. With the help of Figure (1.2), prove that for D_8, it holds true that $cr = r^{-1}c$.

21. Show that $SO_3(\mathbb{R})$ is the group of all rotations about the origin in $\mathbb{R}^3$.

Chapter 2

Group Properties and Group Relations

2.1   Some Set Theory

Groups are sets with extra structure. Specifically, a group
is a set equipped with a multiplication operation, with associativity, with the existence of an identity, and with closure
under multiplication and under inverses. It is therefore an
essential prerequisite to have a firm understanding of basic
set theory. Here, we review a concept from set theory which
will play an important role in what follows.

Definition 2.1. Equivalence relation: Let X be a set and
$x, y, z \in X$. Then an *equivalence relation* on X is a relation
for which the following properties hold:

1. Reflexivity: $x \sim x$ for all $x \in X$.

2. Symmetry: if $x \sim y$ then $y \sim x$.

3. Transitivity: if $x \sim y$ and $y \sim z$ then $x \sim z$.

We read $x \sim y$ as "x is equivalent to y". If $x \sim y$ then we
say "x and y belong to the same *equivalence class*".

Theorem 2.2. *Let X be a set. Then any equivalence relation on X partitions X into a disjoint union of equivalence
classes.*

Proof: Part (a): By reflexivity, each element x of X belongs to some equivalence class, C_x. Part (b): By symmetry,
for any element $y \in X$ for which $x \sim y$, it holds that $y \in C_x$.
By transitivity if $y \in C_z$ then $z \in C_x$, therefore $C_z = C_x$.
i.e. any two equivalence classes are either disjoint of equal.
Together, parts (a) and (b) prove that X is a disjoint union
of equivalence classes. $\boxed{\text{QED}\checkmark}$

Definition 2.3. Injection: An injection is a *one-to-one* mapping between sets. A map σ from a set X to a set Y is injective if for every $x_1, x_2 \in X$ such that $x_1 \neq x_2$, it holds true
that $\sigma(x_1) \neq \sigma(x_2)$.

Example 2.4. The map $f : \mathbb{R} \to \mathbb{R}$ defined by $f(r) = r$ is an injective map. It is clear that if $r_1 \neq r_2$ then $f(r_1) = r_1 \neq r_2 = f(r_2)$.

Example 2.5. The map $f : [0, 2\pi) \to [0, 1]$ defined by $f(\theta) = \cos(\theta)$ is not an injective map. See for instance that $\pi/2 \neq 3\pi/2$ but $\cos(\pi/2) = \cos(3\pi/2) = 0$.

Definition 2.6. Surjection: A surjection is an *onto* map between sets. A map σ from a set X to a set Y is surjective if for any $y \in Y$ there exists an $x \in X$ such that $\sigma(x) = y$.

Example 2.7. The map $\psi : \mathbb{R} \to \mathbb{R}$ defined by $\psi(r) = r/2$ is surjective. To verify this, consider arbitrary $x \in \mathbb{R}$. Then $2x \in \mathbb{R}$ and $\psi(2x) = 2x/2 = x$.

Example 2.8. The map $\psi : \mathbb{Z} \to \mathbb{Z}$ defined by $\psi(k) = 2k$ is not surjective. To verify this, note that there is no $m \in \mathbb{Z}$ such that $\psi(m) = 5$.

Definition 2.9. Bijection: A bijection is an injective surjection. i.e. it is a one-to-one and onto map.

Example 2.10. The map $f : \mathbb{R} \to \mathbb{R}$ defined by $f(r) = r$ is a bijective map. We already showed above that it is injective. It is also clear that for any $r \in \mathbb{R}$, $f(r) = r$. Therefore it is onto.

2.2 Group Attributes

Here we introduce certain fundamental attributes of groups.

Definition 2.11. Subgroup: Let G be a group and H be a subset of G. Then H is a *subgroup* of G if H is itself a group.

For example, Let $\mathbb{Z}$ be the group of integers. Then one of its subgroups is the group of even integers. Of note, the identity is a subgroup of all groups.

Definition 2.12. Order of a group: The order of a group is the number of elements in the group. Given a group G, the order of G is denoted as $|G|$.

For example, the set of symmetries of the square is called the dihedral group of order 8. It is denoted D_8 and it holds that $|D_8| = 8$. The cyclic group of order 11 is denoted $\mathbb{Z}/11$, and it holds that $|\mathbb{Z}/11| = 11$. As we saw in the previous chapter the order of the Klein four group is 4, and as we will see in the coming chapters, it is denoted $\mathbb{Z}/2 \times \mathbb{Z}/2$. Some other groups have infinite order. For example, $|\mathbb{R}^\times| = \infty$, $|\mathbb{R}^+| = \infty$, and $|GL_n(\mathbb{R})| = \infty$ to name a few.

Definition 2.13. Order of an element: The smallest number of times one multiplies the element by itself to get the identity. We denote the order of x by $|x|$.

$$|x| = \min\{n \in \mathbb{N} \text{ such that } x^n = e\}. \qquad (2.1)$$

If G is finite, then $|x|$ is finite. Furthermore, $|x| \leq |G|$.

Remark. If G is a cyclic group, then there exists some $g \in G$ such that $G = \{g, g^2, g^3, \ldots, g^{|g|}\}$. We say g is a generator of G and we write $G = \langle g \rangle$.

Definition 2.14. Generator: A group G has a generator if there exists an element $g \in G$ such that

$$G = \{\ldots, g^{-2}, g^{-1}, e, g, g^2, \ldots\}. \qquad (2.2)$$

The element g is called the *generator* of G, and G is termed *cyclic*. We write $G = \langle g \rangle$.

Example 2.15. $(5\mathbb{Z}, +) = \langle 5 \rangle$.

Example 2.16. $(\mathbb{Z}, +) = \langle 1 \rangle$.

Definition 2.17. Generating set: Let G be a group and S be a subset of G. Then S is a *generating set* of G if every element of G is a product of some elements of S and their inverses. We write $G = \langle S \rangle$. S itself need not be a group, i.e. it need not be closed under addition and it need not even contain the identity.

Remark. Of note, it is always true that $G = \langle G \rangle$.

Remark. If T is a subset of a group G. Then the subgroup generated by T is the smallest subgroup of G containing T. It is the intersection of all subgroups of G containing T.

Definition 2.18. Abelian: An abelian group is one in which all elements commute with each other. G is abelian if for any pair $g, h \in G$, it holds true that $gh = hg$.

Example 2.19. $(\mathbb{R}, +)$ is abelian.

Example 2.20. $(\mathbb{R}, \times)$ is abelian.

Example 2.21. $GL_2(\mathbb{R})$ is not abelian. To see this consider the following. Let

$$\begin{pmatrix} a & b \\ c & d \end{pmatrix} \quad \text{and} \quad \begin{pmatrix} e & f \\ g & h \end{pmatrix} \tag{2.3}$$

be invertible matrices with real entries. Then consider the two possible orders of multiplication,

$$\begin{pmatrix} a & b \\ c & d \end{pmatrix} \begin{pmatrix} e & f \\ g & h \end{pmatrix} = \begin{pmatrix} ae + bg & af + bh \\ ce + dg & cf + dh \end{pmatrix} \tag{2.4}$$

and

$$\begin{pmatrix} e & f \\ g & h \end{pmatrix} \begin{pmatrix} a & b \\ c & d \end{pmatrix} = \begin{pmatrix} ae + cf & eb + fd \\ ag + ch & gb + dh \end{pmatrix}. \tag{2.5}$$

We see that

$$\begin{pmatrix} ae + bg & af + bh \\ ce + dg & cf + dh \end{pmatrix} \neq \begin{pmatrix} ae + cf & eb + fd \\ ag + ch & gb + dh \end{pmatrix}. \tag{2.6}$$

Therefore

$$\begin{pmatrix} a & b \\ c & d \end{pmatrix}\begin{pmatrix} e & f \\ g & h \end{pmatrix} \neq \begin{pmatrix} e & f \\ g & h \end{pmatrix}\begin{pmatrix} a & b \\ c & d \end{pmatrix} \qquad (2.7)$$

in general, because the matrices being multiplied are arbitrary elements of $GL_2(\mathbb{R})$. Complete rigor yet requires that one show the constraints of invertibility and realness do not somehow force equality on the above two products. We can proceed by the simple example –which by itself suffices as a proof of the assertion.

$$\begin{pmatrix} 4 & 3 \\ 3 & 2 \end{pmatrix}\begin{pmatrix} 1 & 2 \\ 3 & 4 \end{pmatrix} = \begin{pmatrix} 13 & 20 \\ 9 & 14 \end{pmatrix}. \qquad (2.8)$$

$$\begin{pmatrix} 1 & 2 \\ 3 & 4 \end{pmatrix}\begin{pmatrix} 4 & 3 \\ 3 & 2 \end{pmatrix} = \begin{pmatrix} 20 & 7 \\ 24 & 17 \end{pmatrix}. \qquad (2.9)$$

And since,

$$\begin{pmatrix} 13 & 20 \\ 9 & 14 \end{pmatrix} \neq \begin{pmatrix} 20 & 7 \\ 24 & 17 \end{pmatrix}, \qquad (2.10)$$

it follows that

$$\begin{pmatrix} 4 & 3 \\ 3 & 2 \end{pmatrix}\begin{pmatrix} 1 & 2 \\ 3 & 4 \end{pmatrix} \neq \begin{pmatrix} 1 & 2 \\ 3 & 4 \end{pmatrix}\begin{pmatrix} 4 & 3 \\ 3 & 2 \end{pmatrix}. \qquad (2.11)$$

Definition 2.22. Center: The *center* of a group is the subset of elements of the group which commute with every element of the group. We denote the center $Z(G)$.

$$Z(G) = \{x \in G \mid xg = gx \text{ for all } g \in G\} \qquad (2.12)$$

Remark. If G is an abelian group, then $Z(G) = G$.

Theorem 2.23. $Z(G)$ *is a subgroup of* G.

Proof: To prove this, we must show that $Z(G)$ satisfies the *group* defining properties. Let $Z(G)$ be the subset of G defined above. By definition the identity commutes with all elements of G, therefore $e \in Z(G)$. Associativity follows directly because G is a group. To show closure under inverse, let $x \in Z(G)$, then $xg = gx$ for all $g \in G$. Taking the inverse of both sides we get $g^{-1}x^{-1} = x^{-1}g^{-1}$ for all $g^{-1} \in G$. Since G is a group and therefore closed under inverses, it follows that $x^{-1}g = gx^{-1}$ for all $g \in G$. Therefore if $x \in Z(G)$ then $x^{-1} \in Z(G)$. To show closure under multiplication, let $x, y \in Z(G)$ and g be an arbitrary element of G. Then

$$xy = yx \tag{2.13}$$
$$yg = gy$$
$$x(yg) = (yg)x$$
$$(xy)g = g(yx)$$
$$(xy)g = g(xy).$$

$\boxed{\text{QED}\checkmark}$

Example 2.24. The center of D_8 is the two element group consisting of the identity and r^2. Of note, r^2 is rotation by 180 degrees.

$$Z(D_8) = \{e, r^2\}. \tag{2.14}$$

Proof: Recall that $D_8 = \{e, r, r^2, r^3, c, cr, cr^2, cr^3\}$. One can check element-wise that no other element is in $Z(D_8)$. For example $r^3c = cr^{-3} = cr \neq cr^3$. More generally, one can verify that for dihedral groups,

$$crc = r^{-1} \tag{2.15}$$
$$cr = r^{-1}c$$

Taking the kth power yields,

$$cr^kc = r^{-k} \tag{2.16}$$
$$cr^k = r^{-k}c$$

If n is even, such as in the square, then $r^{n/2} = r^{-n/2}$. This is because rotating a square 2 turns clockwise is the exact same transformation as rotating the square 2 turns counterclockwise. The "2" here is from $n/2$ where $n = 4$, the number of sides of a square. For any other k such that $1 \leq k < n - 1$ and $k \neq n/2$, it holds true that $r^k \neq r^{-k}$. And for the case where n is odd, $r^{n/2}$ is not a symmetry transformation. Therefore $k = n/2$ where n is even is the only case for which $r^k = r^{-k}$. For n even, the above Equation (2.17) yields,

$$cr^{n/2}c = r^{-n/2} = r^{n/2} \tag{2.17}$$
$$cr^{n/2} = r^{n/2}c$$

Since all rotations commute with each other, the above demonstrated commutation of $r^{n/2}$ with c completes the proof.
QED✓

Definition 2.25. Centralizer: Let G be a group and $x \in G$. Then the centralizer of x in G is the set of $g \in G$ that commute with x. We denote the centralizer of x by $C(x)$.

$$C(x) = \{g \in G \mid xg = gx\} \tag{2.18}$$

Theorem 2.26. *Let G be a group and x be an element of G. Then $C(x)$ is a subgroup of G.*

Proof: Since the identity commutes with all $g \in G$ it holds true that the identity is in $C(x)$. By the definition of the inverse and uniqueness of the identity it holds true that $xx^{-1} = x^{-1}x = e$. Therefore $x^{-1} \in C(x)$. Associativity is inherited trivially from G. To show closure under multiplication, let $g, h \in C(x)$. Then,

$$xg = gx \text{ and } xh = hx \tag{2.19}$$
$$xgh = gxh = ghx$$
$$x(gh) = (gh)x$$
$$gh \in C(x)$$

QED✓

2.3 Group Relations

Definition 2.27. Homomorphism: Let ϕ be a map from a group G to a group H, i.e. $\phi : G \to H$. Then ϕ is a *homomorphism* if for any $x, y \in G$,

$$\phi(xy) = \phi(x)\phi(y). \tag{2.20}$$

Here it is understood that $\phi(x)$ and $\phi(y)$ are elements of H and their product is by the group multiplication with which H is equipped. And the product of x and y is by the group multiplication with which G is equipped.

In effect, a homomorphism is a map that preserves the multiplicative group structure.

Example 2.28. The trivial map is a homomorphism. i.e. the map $I : G \to H$ defined by $I(g) = e_H$ for every $g \in G$. Here, e_H denotes the identity in H.

Proof: For any $x, y \in G$, $I(xy) = e_H = e_H e_H = I(x)I(y)$. QED✓

Example 2.29. Let $\psi : \mathbb{R}^+ \to S^1$ be defined by $\psi(r) = \exp(2\pi i r)$, where $i^2 = -1$, and S^1 is the unit circle group. Then ψ is a homomorphism.

Proof: Let $x, y \in \mathbb{R}$. Then $\psi(x+y) = \exp(2\pi i(x+y)) = \exp(2\pi i x) \exp(2\pi i y) = \psi(x)\psi(y)$. QED✓

Example 2.30. Let $\sigma : \mathbb{Z}^+ \to \{+1, -1\}$ be defined by $\sigma(m) = +1$ if m is even, and $\sigma(m) = -1$ if m is odd. Then σ is a homomorphism. Here we understand that the group operation in the image group is real number multiplication.

Proof: If both m and n are even then $m + n$ is even and $\sigma(m + n) = +1 = (+1)(+1) = \sigma(m)\sigma(n)$. If m is even and n odd, then $m + n$ is odd and $\sigma(m + n) = -1 = (+1)(-1) = \sigma(m)\sigma(n)$. If both are odd then their sum is even and $\sigma(m + n) = +1 = (-1)(-1) = \sigma(m)\sigma(n)$. $\boxed{\text{QED}\checkmark}$

Theorem 2.31. *Let $\phi : G \to H$ be a group homomorphism, and let $x \in G$. Then $|\phi(x)|$ divides $|x|$.*

Proof: Let $|x| = n$. Then $x^n = e$.

$$\phi(e) = \phi(x^n) = (\phi(x))^n \tag{2.21}$$
$$\phi(e)\phi(x) = \phi(x^n)\phi(x) = (\phi(x))^n\phi(x)$$
$$\phi(x) = \phi(x)(\phi(x))^n$$

Left multiplying both sides of the equation by $[\phi(x)]^{-1}$ yields,

$$[\phi(x)]^{-1}\phi(x) = [\phi(x)]^{-1}\phi(x)(\phi(x))^n \tag{2.22}$$
$$e = (\phi(x))^n$$

What the above equation means is that $n = kl$ for some $k, l \in \mathbb{N}$ such that $k = \min\{m \in \mathbb{N} \mid (\phi(x))^m = e\}$. In other words $|\phi(x)| = k$ and $k = n/l$. This technically completes the proof, but to see more note that

$$e = (\phi(x))^k \tag{2.23}$$
$$e = e^l = (\phi(x))^{kl} = (\phi(x))^n$$

The above analysis explicitly demonstrates that $|\phi(x)|$ divides $|x|$. $\boxed{\text{QED}\checkmark}$

Definition 2.32. Kernel: Let G and H be groups and let ϕ be a map from G to H. The kernel of ϕ is the set of elements which ϕ maps onto the identity. We denote the kernel $Ker(\phi)$.

$$Ker(\phi) = \{g \in G \mid \phi(g) = e_H\}, \qquad (2.24)$$

where e_H is the identity in H. One can readily show that $Ker(\phi)$ is a group.

Definition 2.33. Isomorphism: An *isomorphism* is a bijective homomorphism. This means that if ϕ maps a group G to a group H, ϕ is an isomorphism if it is all of the following:

1. A homomorphism

2. One-to-one

3. Onto

We say G is isomorphic to H, and we write $G \simeq H$.

The concept of isomorphism is a powerful one. If two groups are isomorphic then they are essentially the same group, regardless of how differently they are represented. This is because they have the same internal multiplicative structure relating the elements to each other; they have the same number of elements; and each element has one and only one counterpart in the other group — and counterparts behave identically to each other in relation to the other elements. By analogy, an orange fruit is called an orange in English and French, it is called ọsàn in Yoruba, and it is called orenji kajitsu in Japanese. The underlying fruit being described is a real, tangible, edible thing whose existence is independent of the various languages, names, or perspectives by which we choose to describe it. Similarly, two groups are isomorphic if they are the same up to representation. As a result, we are interested in studying isomorphism classes. This provides a deeper insight and understanding of group structure and the relationships between groups.

Theorem 2.34. *Isomorphism is an equivalence relation.*

Proof: We have only to show reflexivity, symmetry, and transitivity. For reflexivity, note that $G \simeq G$ by virtue of the identity map defined by $\phi(g) = g$. This map is obviously an isomorphism. To show symmetry, note that if $G \simeq H$ then there exists an isomorphic map $\phi : G \to H$. Of note, since ϕ is a bijection it is invertible. Therefore there exists a bijection $\phi^{-1} : H \to G$. To show that ϕ^{-1} is a homomorphism, consider $x, y \in G$.

$$\phi(xy) = \phi(x)\phi(y) \qquad (2.25)$$
$$xy = \phi^{-1}(\phi(xy)) = \phi^{-1}(\phi(x)\phi(y))$$
$$xy = \phi^{-1}(\phi(x))\phi^{-1}(\phi(y))$$
$$\phi^{-1}(\phi(x)\phi(y)) = \phi^{-1}(\phi(x))\phi^{-1}(\phi(y))$$

Therefore ϕ^{-1} is an isomorphism, hence $H \simeq G$. Finally to prove transitivity, note that if $G \simeq H \simeq K$ are groups. Then there is an isomorphism ϕ between G and H and an isomorphism ψ between H and K. The composition of isomorphisms is an isomorphism, therefore $\phi \circ \psi$ is an isomorphism between G and K.

Definition 2.35. Automorphism: An automorphism is an isomorphism from a group to itself. If $\phi : G \to G$, then ϕ is an automorphism if it is an isomorphism.

Theorem 2.36. *Let G be a group and $Aut(G)$ the set of automorphisms of G. Then $Aut(G)$ is a group.*

Proof: The group operation is composition. The identity map, i.e. $I(g) = g$ is an isomorphism therefore the identity is in $Aut(G)$. We showed in the above proof that the inverse of an isomorphism is an isomorphism, therefore $Aut(G)$ is closed under inverse. We also showed above that compositions of isomorphisms yield isomorphisms, therefore $Aut(G)$ is closed under multiplication. And finally, since function composition is associative, it follows directly that $Aut(G)$ has the associativity property. $\boxed{\text{QED}\checkmark}$

2.4   Exercises

1. Show that the composition of isomorphisms is an isomorphism.

2. Let ϕ be a group homomorphism. Show that $Ker(\phi)$ is a group.

3. Show that $GL_3(\mathbb{R})$ is not abelian.

4. Show that $GL_n(\mathbb{R})$ is not abelian for any $n \in \mathbb{N}$.

5. Show that $Z(D_8) = \{e, r^2\}$.

6. Prove that $crc = r^{-1}$ where c and r are respectively reflection and rotation elements of D_{2n}.

7. Find the center of D_{2n} when n is even.

8. Find the center of D_{2n} when n is odd.

9. Find the center of $GL_2(\mathbb{R})$.

10. Let G be a group. Prove that $Z(G)$ is a subgroup of G.

11. Let G be a group and x an element of G. Prove that $C(x)$ is a subgroup of G.

12. Prove that $Z(G) \subset C(x)$ for all $x \in G$.

13. Consider an equivalence relation defined by: $x \sim y$ if $y = gxg^{-1}$ for some $g \in G$. Prove that if $C_x = \{x\}$ then $x \in Z(G)$.

14. Show that the map $\pi_g : G \to G$ given by $\pi_g(x) = gxg^{-1}$ is an automorphism.

15. Show that if $\sigma : X \to Y$ is an isomorphism, then $|x| = |\sigma(x)|$.

16. Let p be a prime and x be a non-trivial natural number such that $x \in \mathbb{Z}/p$. Show that $\mathbb{Z}/p = \langle x \rangle$.

17. Describe the automorphisms of $(\mathbb{Z}/6, +)$.

18. Describe the automorphisms of $(\mathbb{Z}/7, +)$.

19. Show that the multiplicative group $\{+1, -1\}$ is isomorphic to $(\mathbb{Z}/2, +)$.

20. Let $\psi : G \to H$ be a group homomorphism.

   a) Show that for any $g \in G$, $|\psi(g)|$ divides $|g|$.

   b) Does $|\psi(g)|$ necessarily equal $|g|$? Why or why not?

Chapter 3

Quotients, Cosets, and Lagrange's Theorem

3.1 Normality

Definition 3.1. Let $N \subset G$ be groups, then N is said to be *normal* in G if for every $g \in G$ and every $n \in N$ it holds that $gng^{-1} \in N$. Stated differently, $gNg^{-1} = N$. We write $N \triangleleft G$.

Example 3.2. The subgroup of rotations is normal in the dihedral group. i.e. let $N_{rot} := \{e, r, r^2, \ldots, r^{n-1}\}$, then

$$N_{rot} \triangleleft D_{2n}. \tag{3.1}$$

Proof:

$$crc^{-1} = crc = r^{-1} \in N_{rot} \tag{3.2}$$
$$(crc)^k = cr^k c = r^{-k} \in N_{rot}$$
$$r^l r^k r^{-l} = r^k \in N_{rot}$$

Therefore $d(N_{rot})d^{-1} = N_{rot}$ for all $d \in D_{2n}$. $\boxed{\text{QED}\checkmark}$

Definition 3.3. A group is called *simple* if its only normal subgroups are itself and the identity.

The notion of *simple* connotes an extreme of non-abelianness. Note that every subgroup of an abelian group is normal. Therefore in this sense simple groups and abelian groups lie on opposite extremes of the 'commutativity spectrum'. The classification of finite simple groups is largely regarded as a pinnacle triumph of collaborative mathematics over this past century. The classification is thought to be largely complete, but undoubtedly some finishing touches remain to be added, such as that of Koichiro Harada and Ron Solomon as recently as 2008.[6] As we will see in the final chapter, via the fundamental theorem of finite abelian groups, we also have traction on the classification of abelian group. Therefore both ends of the 'commutativity spectrum' can be said to have been characterized.

3.2 Quotients and Cosets

The notion of quotients applies to groups and other algebraic structures. Given a group G and a subgroup H, we can construct a structure called a coset space of G 'quotiented' by H. The coset space is denoted G/H and is of the following form:

$$G/H = \{gH \mid g \in G\} \tag{3.3}$$

In particular, the above is called the *left coset space*. This is in contrast to the *right coset space* which is given by,

$$H\backslash G = \{Hg \mid g \in G\} \tag{3.4}$$

Definition 3.4. Let $H \subset G$ be groups. Then $[G : H]$ is called the *index of H in G* and is given by,

$$[G : H] = |G/H| = \frac{|G|}{|H|} \tag{3.5}$$

Proposition 3.5. *Quotienting G by H — i.e. forming G/H — induces an equivalence relation in which each coset gH is an equivalence class.*

   Proof: Let us define $x \sim y$ to mean that x belongs to the coset yH. Our goal then is to show that this indeed is an equivalence relation. To see reflexivity, note that since $x \in xH$, it follows from our definition that $x \sim x$. To show symmetry, let $x \sim y$ and therefore $x \in yH$. Then $x = yh$ for some $h \in H$. It follows that $y = xh^{-1} \in xH$. Therefore $y \sim x$. And finally, we need to show transitivity. Let $x \sim y$ and $y \sim z$, i.e. $x \in yH$ and $y \in zH$. Then $x = yh$ for some $h \in H$ and $y = zh'$ for some $h' \in H$. Substituting we see that $x = zh'h \in zH$. Therefore $x \sim z$. $\boxed{\text{QED}\checkmark}$

   A coset space G/H is a partitioning of a group into cosets xH, which are sets of size $|H|$ as illustrated in figure (3.1).

The denominator group H contains the identity and is privileged in that regard. And H is the only one of the cosets that is a group.

Figure 3.1: Cosets

Theorem 3.6. *Let G be a group and N a subgroup of G. If N is normal in G then G/N is itself a group.*

Proof: Let $N \triangleleft G$, and consider any two elements of G/N call them gN and hN. By the definition of *normal* $gN = Ng$ and $hN = Nh$. To show that G/N is a group, we show group properties in turn. G/N has an identity, i.e. $eN = N$. Each element gN has an inverse: $(gN)^{-1} = Ng^{-1} = g^{-1}N$. Observe that $(gN)(g^{-1}N) = gg^{-1}N = N$. To verify closure under multiplication, note that $(gN)(hN) = ghN \in G/N$. To verify associativity, note that $gN(hNkN) = (gN)(hkN) = ghkN = (gNhN)(kN)$. $\boxed{\text{QED}\checkmark}$

Theorem 3.7. *If N is normal in G, then the left coset space G/N equals the right coset space $N\backslash G$*

Proof: By the definition of *normal*, $gN = Ng$ for every $g \in G$. $\boxed{\text{QED}\checkmark}$

Theorem 3.8. *Index Two Theorem: Let $H \subset G$ be groups such that $|H| = |G|/2$. Then $H \triangleleft G$.*

Proof: Consider the left coset space G/H. The coset $eH = H$ constitutes half of the elements of G while the coset

gH for any $g \notin H$ constitute the remaining elements. For the right coset space $H\backslash G$ we have cosets $He = H$ and Hg for $g \notin G$. The coset H and its complement in G are the two elements of both coset spaces and must therefore be equal. It follows that for every $g \in G$,

$$gH = Hg \tag{3.6}$$
$$gHg^{-1} = H$$

$\boxed{\text{QED}\checkmark}$

3.3 Lagrange's Theorem

Theorem 3.9. *Let $S \subset G$ be groups. Then the order of S divides the order of G.*

Proof: Quotienting G by S generates an equivalence relation which partitions the elements of G into the coset space G/S. This is a disjoint union of sets each of size $|S|$. Therefore,

$$|G| = |S||G/S|. \tag{3.7}$$

Therefore,

$$|S| = \frac{|G|}{|G/S|}. \tag{3.8}$$

Interestingly, though this theorem is clear and its proof is straight forward and simple as above, it is of great importance. It features recurrently throughout the remainder of this book and throughout all of algebra in general.

3.4 Correspondence Theorem

Theorem: Let N and G be groups such that $N \triangleleft G$, and let K be a subgroup of the quotient group G/N. Then there exists a subgroup J of G such that,

$$J = \{xn \in G \mid n \in N \text{ and } xN \in K\}. \qquad (3.9)$$

This is also known as the lattice theorem.

Proof: Here we need to show that the elements of J satisfy the requirements for being a group. The elements of K are of the form xN where $x \in G$. The identity element is N, and the other elements are of the form zN where $z \notin N$ is a representative element from the zN coset. Since K is a group, the elements of K respect the properties of group such as closure under multiplication and closure under inverse. Proceeding with the proof that J is a group, it is trivial to show associativity and existence of identity. Specifically, N is a subset of J and contains the identity, while associativity is directly inherited from G. We next show closure under multiplication and under inverses:

Closure under multiplication. Consider arbitrary $y, z \in J$. We consider three cases. Case i) If $y, z \in N$ we are done, since N is a subgroup of G and a subset of J. Case ii) If $y \in N$, but $z \notin N$. For clarity, relabel $y = n \in N$. Since $z \in J$ it follows that $z = z'n'$ for some $n' \in N$ and some $z'N \in K$. Therefore $yz = nz = nz'n' = z'(n''n')$, for some $n'' \in N$. We used $nz' = z'n''$ which follows from the normalcy of N. We have shown that $yz = z'(n''n')$ where $n''n' \in N$ and $z'N \in K$. Hence we have shown that $yz \in J$. Since our group is not necessarily abelian, we are not yet done with this case. We still need to show $zy \in J$. Note $zy = zn = z'(n'n)$ and $z'N \in K$, hence $zy \in J$. Case iii) If $z, y \notin N$, then since K is a group and therefore closed under multiplication, $zyN \in K$ and therefore $zy \in J$. Similarly, $yz \in J$.

Closure under inverses. Consider arbitrary $y \in J$. Case i) If $y \in N$ then so is y^{-1}. This is true since N is a subgroup of G and a subset of J. Case ii) If $y \notin N$, then since N a group, $y^{-1} \notin N$ either. In K, closure under inverse implies there exists a $zN \in K$ — $z \in G$, $z \notin N$ — such that $yz \in N$. This is because $(yN)(zN) = yzN = N$ only requires $yz \in N$. yz need not equal the identity in G. i.e. z need not equal y^{-1}. Therefore we must still show existence of $y^{-1} \in J$ directly. Since $yz \in N$, $yz = n$ for some $n \in N$. Therefore $z = y^{-1}n$, and $zN = y^{-1}nN = y^{-1}N \in K$. Therefore $y^{-1} \cdot e_N = y^{-1} \in J$. This concludes the proof. $\boxed{\text{QED}\checkmark}$.

Commentary: *Lysing the N-membrane.* The elements of K each have cardinality of N when viewed in the granularity of G. The elements of K are more course grained than those of G, and specifically they are $|N|$ times larger than those of G because each element of K contains $|N|$ elements of G. This theorem states that if we lyse the N-membranes surrounding each element of K and spill their contents, the resulting collection of such contents form a subgroup of G. In the proof, we proceeded to show that the resultant contents after lysing satisfies group properties.

3.5 Exercises

1. Does the set of natural numbers under addition form a group? Why or why not?

2. Verify that the set of even integers under addition form a group.

3. Does the set of odd integers under addition form a group? Why or why not?

4. Show that every subgroup of an abelian group is normal.

5. Let $K \lhd G$ and $H \lhd G$. Show that $K \cap H \lhd G$.

6. Let $N, H \subset G$ be groups such that $N \lhd G$. And such that $[G : N]$ and $|H|$ are relatively prime. Prove that $H \subset N$.

7. Let $N \lhd G$ be groups. Prove that G/N is a group.

8. Let K be a subgroup of the quotient group G/H. Show that there exists a subgroup J of G such that,

$$J = \{xh \in G | h \in H \text{ and } xH \in K\}$$

9. Let $GL_n(\mathbb{R})$ denote the generalized linear group, i.e. the group of invertible $n \times n$ matrices with real entries. Consider the map $\phi : GL_n(\mathbb{R}) \to \mathbb{R}^\times$ defined as

$$\phi(M) = det(M),$$

where $\mathbb{R}^\times$ denotes the group of real numbers under multiplication and $det(M)$ denotes the determinant of M.

   a) Show that ϕ is a homomorphism.

   b) Show that ϕ is onto.

10. Let $SL_n(\mathbb{R})$ denote the special linear group, i.e. the subgroup of $GL_n(\mathbb{R})$ with determinant of 1. Show that $SL_n(\mathbb{R}) \lhd GL_n(\mathbb{R})$.

11. Show that $GL_n(\mathbb{R})/SL_n(\mathbb{R}) \simeq \mathbb{R}^\times$.

Chapter 4

The Commutator Subgroup

The commutator subgroup is a special subgroup which provides a way to grade the extent to which a group is abelian. Yes indeed a group is either abelian or not abelian. However of the groups which are not abelian, it is useful to know how close the are to abelian. The commutator subgroup which we will define below gives us a good precise handle on this notion.

Definition 4.1. Let G be a group and $g, h \in G$, then the *commutator* of g and h denoted $[g, h]$ is given by,

$$[g, h] = g^{-1}h^{-1}gh. \tag{4.1}$$

Definition 4.2. The commutator subgroup of G denoted $[G, G]$ is the subgroup generated by the set of commutators of all pairs of elements of G.

$$[G, G] = \tag{4.2}$$
$$\langle \{[g, h] \mid g, h \in G\} \rangle =$$
$$\langle \{g^{-1}h^{-1}gh \mid g, h \in G\} \rangle.$$

Theorem 4.3. *The commutator subgroup $[G, G]$ is normal in G.*

Proof: Consider an arbitrary element x of $[G, G]$. From the definition, x is of the form,

$$x = a_1^{-1}b_1^{-1}a_1b_1a_2^{-1}b_2^{-1}a_2b_2 \ldots a_n^{-1}b_n^{-1}a_nb_n \tag{4.3}$$

where $a_i, b_i \in G$. Consider gxg^{-1} for an arbitrary element $g \in G$.

$$gxg^{-1} = \qquad (4.4)$$
$$ga_1^{-1}b_1^{-1}a_1b_1a_2^{-1}b_2^{-1}a_2b_2\ldots a_n^{-1}b_n^{-1}a_nb_ng^{-1} =$$
$$(ga_1^{-1}g^{-1})(gb_1^{-1}g^{-1})(ga_1g^{-1})(gb_1g^{-1})\ldots$$
$$\ldots(ga_n^{-1}g^{-1})(gb_n^{-1}g^{-1})(ga_ng^{-1})(gb_ng^{-1}) =$$
$$\hat{a_1}^{-1}\hat{b_1}^{-1}\hat{a_1}\hat{b_1}\ldots\hat{a_n}^{-1}\hat{b_n}^{-1}\hat{a_n}\hat{b_n} =$$
$$[\hat{a_1},\hat{b_1}]\ldots[\hat{a_n}\hat{b_n}] \in [G,G].$$

$\boxed{\text{QED}\checkmark}$

In the third step above we inserted $e = gg^{-1}$ between terms. And in the fifth step, for clarity sake, we relabeled by $\hat{a_i} = ga_ig^{-1}$.

Theorem 4.4. *If G is abelian, then $[G,G] = \{e\}$.*

Proof: If G is abelian, then for every pair $g, h \in G$ it holds that,

$$[g,h] = \qquad (4.5)$$
$$g^{-1}h^{-1}gh =$$
$$g^{-1}gh^{-1}h =$$
$$e \cdot e =$$
$$e.$$

$\boxed{\text{QED}\checkmark}$

Therefore when G is abelian, $[G,G]$ is the smallest possible group, $\{e\}$. In the same vein, we will see that the less abelian G is, the larger $[G,G]$ will be. This notion is embodied in the following theorem:

Theorem 4.5. *Let G be a group and N be a normal subgroup of G. Then G/N is abelian if and only if $[G,G] \subseteq N$. In particular, $G/[G,G]$ is always abelian and hence is called the abelianization of G.*

Proof: Assume G/N is abelian. Then for any pair of elements $gN, hN \in G/N$, it holds that,

$$(gN)(hN) = (hN)(gN)$$

Therefore,

$$ghN = hgN \qquad (4.6)$$
$$(hg)^{-1}ghN = N$$
$$g^{-1}h^{-1}ghN = N$$
$$g^{-1}h^{-1}gh \in N.$$

And since g and h are arbitrary elements of G, it follows that $g^{-1}h^{-1}gh$ is an arbitrary element of $[G, G]$; therefore

$$[G, G] \subseteq N.$$

For the other direction we simply work backwards. Assume $[G, G] \subseteq N$. Then for every pair $g, h \in G$, it holds that,

$$g^{-1}h^{-1}gh \in N \qquad (4.7)$$
$$g^{-1}h^{-1}ghN = N$$
$$(hg)^{-1}ghN = N$$
$$(hg)^{-1}Ngh = N$$
$$Ngh = hgN$$
$$NNgh = hgNN$$
$$(gN)(hN) = (hN)(gN).$$

$$\boxed{\text{QED}\checkmark}$$

Definition 4.6. Solvability: A group G is called *solvable* if it contains a sequence of subgroups $\{e\} = H_0 \subseteq H_1 \subseteq H_2 \subseteq \ldots \subseteq H_n = G$ such that

$$\{e\} = H_0 \lhd H_1 \lhd H_2 \lhd \ldots \lhd H_n = G$$

and for each i,

$$H_{i+1}/H_i \text{ is abelian.}$$

Definition 4.7. Derived series: Given a group G we can derive a series by the following designations:

$$G := G^{(0)} \tag{4.8}$$
$$[G, G] := G^{(1)}$$
$$[G^{(1)}, G^{(1)}] := G^{(2)}$$
$$\vdots$$
$$[G^{(s-1)}, G^{(s-1)}] := G^{(s)}$$

From Theorem (4.3) which states that the commutator subgroup is normal in the group, we know that

$$G^{(i)} \lhd G^{(i-1)} \tag{4.9}$$

for every i. Furthermore, from Theorem (4.5) on the abelianization of a group, we know that

$$G^{(i-1)}/G^{(i)} \text{ is abelian for every } i. \tag{4.10}$$

Examining our derived series above, it is notable that it appears to conform to the requirements for solvability. On that note we introduce the following important theorem about the solvability of finite groups.

Theorem 4.8. *A finite group G is solvable if and only if its derived series terminates in the identity.*

$$G \text{ is solvable} \iff G^{(j)} = \{e\} \text{ for some } j. \tag{4.11}$$

Proof: Part (a): Assume the derived series of a group G terminates in the identity. Then $\{e\} = G^{(j)}$ for some j. We know that $G^{(j)} \lhd G^{(j-1)}$ and that $G^{(j-1)}/G^{(j)}$ is abelian. By process of extension, we have,

$$\{e\} = G^{(j)} \lhd G^{(j-1)} \lhd \cdots \lhd G^{(0)} = G \qquad (4.12)$$

such that

$$G^{(k-1)}/G^{(k)} \qquad (4.13)$$

is abelian for $1 \leq k \leq j$.

Part (b): Assume a group G is solvable. Then there exists a sequence of subgroups $\{e\} = H_0 \subseteq H_1 \subseteq H_2 \subseteq \ldots \subseteq H_m = G$ such that

$$\{e\} = H_0 \lhd H_1 \lhd H_2 \lhd \ldots \lhd H_m = G \qquad (4.14)$$

and for each i,

$$H_{i+1}/H_i \text{ is abelian.} \qquad (4.15)$$

.

Lemma 4.9. *Let U and V be groups such that $U \subset V$. Then $[U, U] \subset [V, V]$. i.e. $U^{(1)} \subset V^{(1)}$.*

Proof of Lemma: For every pair $u_1, u_2 \in U$, it holds that $u_1, u_2 \in V$, and therefore $u_1^{-1} u_2^{-1} u_1 u_2 = [u_1, u_2] \in V$. Therefore $[U, U] \subseteq [V, V] \subseteq V$. $\boxed{\text{QED}\checkmark}$

Since G/H_{m-1} is abelian, from the abelianization theorem (Thm 4.5) it follows that,

$$[G, G] = G^{(1)} \subseteq H_{m-1}. \qquad (4.16)$$

In turn, since H_{m-1}/H_{m-2} is abelian, it holds that

$$[H_{m-1}, H_{m-1}] \subseteq H_{m-2} \qquad (4.17)$$

and from Lemma (4.9) we have that,

$$G^{(2)} = [G^{(1)}, G^{(1)}] \subseteq [H_{m-1}, H_{m-1}] \subseteq H_{m-2}. \qquad (4.18)$$

By the same reasoning it follows inductively that for any j such that $1 \leq j \leq m$,

$$G^{(j)} = \left[G^{(j-1)}, G^{(j-1)}\right] \subseteq [H_{m-j+1}, H_{m-j+1}] \subseteq H_{m-j}. \qquad (4.19)$$

For $j = m$ we have,

$$G^{(m)} = \left[G^{(m-1)}, G^{(m-1)}\right] \subseteq [H_{m-m+1}, H_{m-m+1}] \subseteq H_{m-m}. \qquad (4.20)$$

This yields,

$$G^{(m)} = \left[G^{(m-1)}, G^{(m-1)}\right] \subseteq [H_1, H_1] \subseteq H_0 = \{e\}. \qquad (4.21)$$

And therefore,

$$G^{(m)} = \{e\}. \qquad (4.22)$$

$\boxed{\text{QED}\checkmark}$

4.1 Exercises

1. Show that if a group is abelian, then its commutator subgroup is $\{e\}$.

2. Let G be a group and $g, h \in G$. Show that $[g, h]^{-1} = [h, g]$.

3. Let $\pi_g(x) = g^{-1}xg$. This is called conjugation by g. Show that $\pi_y(x) = x[x, y]$.

4. Show that $\pi_g([x, y]) = [\pi_g(x), \pi_g(y)]$.

5. Let G be a group and S be a subset of G. Prove that the subgroup generated by S is the smallest subgroup of G containing S.

6. Let G be a group and $[G, G]$ be the commutator subgroup of G. Show that $[G, G] \triangleleft G$.

7. Let G be a group, and N be a normal subgroup of G. Show that G/N is abelian if and only if $[G, G] \subseteq N$.

8. Let G be a group. Find the center of $G/[G, G]$.

9. Let G be a group. Find the commutator subgroup of $G/[G, G]$.

10. Prove that a group G is solvable if and only if its derived sequence terminates in the trivial group.

11. Let $H \subseteq G$ be groups. Show that if H is solvable if G is solvable.

Chapter 5

The Isomorphism Theorems

Emmy Noether (1882-1935): *The isomorphism theorems are a set of theorems which describe certain homomorphic relationships between algebraic structures and their quotients and substructures. A number of mathematicians including Emmy Noether and Richard Dedekind contributed to their discovery. Emmy Noether was a leading mathematician whose contributions to algebra, number theory, and physics are far reaching. She also discovered the Noether theorem which features centrally in modern physics and states that for every symmetry there is an associated conserved quantity. Her professional research career began at the University of Göttingen after which she moved to Moscow State University. In 1933 she moved to Bryn Mawr College in the United States where she died and is buried.*

5.1   The First Isomorphism Theorem

Theorem 5.1. *Let $\phi : G \to H$ be a homomorphism between groups G and H. Then,*

1. $Ker(\phi) \lhd G$

2. $G/Ker(\phi)$ *is isomorphic to* $Image(\phi)$.

Proof of 1.: $Ker(\phi) = \{g \in G \text{ such that } \phi(g) = e_H\}$. Therefore, given any $g \in G$, consider $gKer(\phi)g^{-1}$. ϕ maps this as follows,

$$\phi(gKer(\phi)g^{-1}) = \qquad (5.1)$$
$$\phi(g)\phi(Ker(\phi))\phi(g^{-1}) =$$
$$\phi(g)e_H\phi(g^{-1}) =$$
$$\phi(g)\phi(g^{-1}) =$$
$$\phi(gg^{-1}) =$$
$$\phi(e_G) =$$
$$e_H.$$

Therefore, for any $g \in G$,

$$gKer(\phi)g^{-1} \in Ker(\phi) \qquad (5.2)$$
$$\text{i.e. } Ker(\phi) \lhd G$$

QED✓

Proof of 2. To show that $G/Ker(\phi) \simeq Image(\phi)$, we will provide an isomorphism ψ,

$$\psi : G/Ker(\phi) \to Image(\phi) \qquad (5.3)$$

such that,

$$\psi(gKer(\phi)) = \phi(g). \qquad (5.4)$$

To show that ψ is an isomorphism, we show in turn that it is:

1. A homomorphism,

2. A surjection (onto), and

3. An injection (one-to-one).

To show ψ is a homomorphism, consider,

$$\psi(gKer(\phi)hKer(\phi)) \tag{5.5}$$

for some $g, h \in G$. By normality of $Ker(\phi)$ which we have shown,

$$Ker(\phi)h = hKer(\phi). \tag{5.6}$$

And by the homomorphism property of ϕ,

$$\phi(gh) = \phi(g)\phi(h). \tag{5.7}$$

Therefore,

$$\begin{aligned}
\psi\left(gKer(\phi)hKer(\phi)\right) &= \tag{5.8}\\
\psi(ghKer(\phi)Ker(\phi)) &=\\
\psi(ghKer(\phi)) &=\\
\phi(gh) &=\\
\phi(g)\phi(h) &=\\
\psi\left(gKer(\phi)\right)\psi\left(hKer(\psi)\right).&
\end{aligned}$$

The above shows that,

$$\psi\left(gKer(\phi) \cdot hKer(\phi)\right) = \psi\left(gKer(\phi)\right) \cdot \psi\left(hKer(\psi)\right), \tag{5.9}$$

i.e. ψ is a homomorphism. Next we proceed to show that ψ is onto. i.e. that given any $h \in Image(\phi)$ there exists an element K of $G/Ker(\phi)$ such that $\psi(K) = h$. By definition of *image* there exists some $g \in G$ such that $\phi(g) = h$. In turn, by our definition of ψ it follows that,

$$h \in Image(\phi) \Rightarrow \tag{5.10}$$
$$\exists\, g \in G \text{ such that } h = \phi(g)$$
$$= \psi(gKer(\phi)).$$

Therefore ψ maps onto $Image(\phi)$. Finally we must show that ψ is one-to-one. i.e. if $gKer(\phi) \neq hKer(\phi)$, then $\psi(gKer(\phi)) \neq \psi(hKer(\phi))$. It suffices to show that if $gKer(\phi) \neq hKer(\phi)$ then $\phi(g) \neq \phi(h)$. Since the partition of G into cosets is an equivalence relation, two cosets are either identical or disjoint. Therefore if $gKer(\phi) \neq hKer(\phi)$, they have no elements in common. Hence, $g \neq h$. Furthermore, $g^{-1}h \notin Ker(\phi)$, else $gg^{-1}h \in gKer(\phi)$ and hence $h \in gKer(\phi)$. This would mean both g and h are in the same coset which is a contradiction.

$$g^{-1}h \notin Ker(\phi) \Rightarrow \tag{5.11}$$
$$\phi(g^{-1}h) \neq e \Rightarrow$$
$$\phi(g^{-1})\phi(h) \neq e \Rightarrow$$
$$[\phi(g)]^{-1}\phi(h) \neq e \Rightarrow$$
$$\phi(h) \neq \phi(g)$$

QED✓

Corollary 5.2. *Let $\phi : G \to H$ be an onto homomorphism. Then ϕ is injective if and only if the kernel of ϕ is trivial.*

Proof of part (a): Let $Ker(\phi) = \{e\}$. Then using the first isomorphism theorem we see that,

$$Ker(\phi) \simeq \{e\}, \qquad\qquad (5.12)$$
$$G/Ker(\phi) \simeq H,$$
$$G/Ker(\phi) \simeq G/\{e\},$$
$$G \simeq G/\{e\},$$
$$\therefore G \simeq H.$$

Therefore by the definition of isomorphism, ϕ is one-to-one. $\boxed{\text{QED}\checkmark}$

Proof of part (b): Let ϕ be one-to-one. It is therefore an onto one-to-one homomorphism. In other words, ϕ is an isomorphism between G and H. Again using the isomorphism theorem we see that,

$$G \simeq H, \qquad\qquad (5.13)$$
$$G/ker(\phi) \simeq H,$$
$$G \simeq G/ker(\phi),$$
$$\therefore \ ker(\phi) \simeq \{e\}.$$

$\boxed{\text{QED}\checkmark}$

Theorem 5.3. *Let $\phi : G \rightarrow H$ be a group isomorphism. Then for any $g \in G$, $|\phi(g)| = |g|$.*

Proof: Recall from Theorem (2.31) that $|\phi(g)|$ divides $|g|$. And recall from Corollary (5.2) that $Ker(\phi) = \{e\}$. Let $|\phi(g)| = m$ and $|g| = n$. By way of contradiction, assume $|\phi(g)| \neq |g|$. Then $m < n$.

$$e_H = (\phi(g))^m = \phi(g^m) \qquad\qquad (5.14)$$
$$g^m = e_G$$
$$m < |g| = n$$

The above constitutes a contradiction, since by definition $|g| = n$ is the smallest power of g that equals the identity.

$\boxed{\text{QED}\checkmark}$

Examples of the First Isomorphism Theorem Applied

Example 5.4. $\mathbb{R}^+/\mathbb{Z}^+ \simeq S^1$,

where $\mathbb{R}^+$ is the group of reals with addition as group operation, $\mathbb{Z}^+$ is the group of the integers with addition as group operation, and S^1 is the zero-centered unit circle in the complex plane given by

$$S^1 = \{z \in \mathbb{C} \text{ such that } \|z\| = 1\} \qquad (5.15)$$
$$= \{e^{2\pi i\theta} \text{ where } 0 \le \theta < 1\}.$$

Proof: Consider the map,

$$\phi : \mathbb{R}^+ \to S^1 \qquad (5.16)$$

given by,

$$\phi(r) = \exp\left(2\pi i r\right). \qquad (5.17)$$

We claim that ϕ is a homomorphism. Proof:

$$\phi(r_1 + r_2) = \qquad (5.18)$$
$$\exp\left(2\pi i(r_1 + r_2)\right) =$$
$$\exp\left(2\pi i r_1\right)\exp\left(2\pi i r_2\right) =$$
$$\phi(r_1)\phi(r_2).$$

We claim that ϕ is onto. Proof: For an arbitrary element $\exp(2\pi i\theta)$ of S^1, note that $\phi(\theta) = \exp(2\pi i\theta)$. Therefore ϕ is onto. The kernel of ϕ is $\mathbb{Z}^+$.

$$Ker(\phi) = \mathbb{Z}^+ \tag{5.19}$$

therefore by the first isomorphism theorem,

$$\mathbb{R}^+/Ker(\phi) \simeq S^1 \tag{5.20}$$
$$\mathbb{R}^+/\mathbb{Z}^+ \simeq S^1$$

$\boxed{\text{QED}\checkmark}$

Example 5.5. $GL_n(\mathbb{R})/SL_n(\mathbb{R}) \simeq \mathbb{R}^\times$,

where $GL_n(\mathbb{R})$ denotes the generalized linear group, i.e. the group of invertible $n \times n$ matrices with real entries; $SL_n(\mathbb{R})$ is its subgroup consisting of elements with determinant 1; and $\mathbb{R}^\times$ are the reals under multiplication.

Proof: Consider the map $\phi : GL_n(\mathbb{R}) \to \mathbb{R}^\times$ given by $\phi(A) = det(A)$. We claim ϕ is a homomorphism. Proof:

$$\phi(AB) = \tag{5.21}$$
$$det(AB) =$$
$$det(A)det(B) =$$
$$\phi(A)\phi(B).$$

We claim that ϕ is onto. Proof: Let r be an arbitrary element of $\mathbb{R}^\times$. And choose $A \in GL_n(\mathbb{R})$ as follows,

$$A = \begin{pmatrix} r & 0 & 0 \\ 0 & 1 & 0 \\ 0 & 0 & 1 \end{pmatrix}. \tag{5.22}$$

It follows that,

$$det(A) = r. \tag{5.23}$$

Therefore ϕ is onto. The kernel of ϕ is the subgroup of matrices with determinant 1. This is exactly the definition of $SL_n(\mathbb{R})$. Therefore by the first isomorphism theorem we have,

$$GL_n(\mathbb{R})/ker(\phi) \simeq \mathbb{R}^\times \qquad (5.24)$$
$$GL_n(\mathbb{R})/SL_n(\mathbb{R}) \simeq \mathbb{R}^\times.$$

QED✓

5.2   Second Isomorphism Theorem

Theorem 5.6. *Let $N, H \subset G$ be groups such that $N \triangleleft G$. Then*

1. *HN is a subgroup of G,*

2. *$H \cap N \triangleleft H$, and*

3. *$HN/H \simeq H/H \cap N$.*

Proof of Part 1. To show HN is a subgroup of G, we show closure under multiplication, inverse, associativity, and identity. Let $h_1, h_2 \in H$ and $n_1, n_2 \in N$. Consider $h_1 n_1, h_2 n_2 \in HN$. Then since $N \triangleleft G$, it follows that $h_2^{-1} n_1 h_2 \in N$. i.e. $h_2^{-1} n_1 h_2 = n_1'$ for some $n_1' \in N$. Therefore $n_1 h_2 = h_2 n_1'$. And therefore, $h_1 n_1 h_2 n_2 = h_1 h_2 n_1' n_2 \in HN$. Hence HN is closed under multiplication. For inverses, consider $hn \in HN$, then $(hn)^{-1} = n^{-1} h^{-1} \in HN$ because N is normal, i.e. because $HN = NH$. Therefore HN is closed under inverses. For associativity, consider,

$$h_1 n_1 (h_2 n_2 h_3 n_3) = \tag{5.25}$$
$$h_1 n_1 (h_2 h_3 \hat{n}_2 n_3) =$$
$$h_1 h_2 h_3 \hat{n}_1 \hat{n}_2 n_3$$

where

$$\hat{n}_2 = h_3^{-1} n_2 h_3 \tag{5.26}$$

and

$$\hat{n}_1 = (h_2 h_3)^{-1} n_1 (h_2 h_3). \tag{5.27}$$

Now consider,

$$(h_1 n_1 h_2 n_2) h_3 n_3 = \tag{5.28}$$
$$(h_1 h_2 n_1' n_2) h_3 n_3 =$$
$$h_1 h_2 h_3 n'' n_3$$

where

$$n_1' = h_2^{-1} n_1 h_2 \tag{5.29}$$

and

$$n'' = \tag{5.30}$$
$$h_3^{-1} n_1' n_2 h_3 =$$
$$h_3^{-1} h_2^{-1} n_1 h_2 n_2 h_3 =$$
$$(h_2 h_3)^{-1} n_1 h_2 n_2 h_3 =$$
$$\hat{n}_1 \hat{n}_2.$$

Substituting $n'' = \hat{n}_1 \hat{n}_2$ above, we get,

$$h_1 n_1 (h_2 n_2 h_3 n_3) = (h_1 n_1 h_2 n_2) h_3 n_3. \tag{5.31}$$

This proves associativity. Finally, it is clear that $e \in HN$.
$\boxed{\text{QED}\checkmark}$

Proof of Part 2. We seek to show that if $N \lhd G$ and H is a subgroup of G, then $H \cap N \lhd H$. From normality of N in G, it follows that for any $n \in N$ and for any $g \in G$, $gng^{-1} \in N$. Therefore for any $x \in H \cap N$ and for any $h \in H$, $hxh^{-1} \in N$. Also, since h, x, $h^{-1} \in H$, by closure under group multiplication it follows that $hxh^{-1} \in H$. Therefore hxh^{-1} is in both H and N, and hence is in $H \cap N$. This shows that $H \cap N \lhd H$. $\boxed{\text{QED}\checkmark}$

Proof of Part 3. Here we seek to show that if $N \lhd G$ and H is a subgroup of G, then $HN/H \simeq H/H \cap N$. Consider the map ϕ,

$$\phi : H \to HN/N \qquad (5.32)$$

defined by,

$$\phi(h) = hN. \qquad (5.33)$$

We claim that ϕ is a homomorphism. To see this consider $\phi(hk)$ for $h, k \in H$.

$$\phi(hk) = hkN \qquad (5.34)$$
$$= hNkN = \phi(h)\phi(k). \quad \checkmark$$

The above $hkN = hkNN = hNkN$ follows from the normalcy of N. We claim that ϕ is onto. To see this, consider an arbitrary element $hnN \in HN/N$. By definition, $\phi(h) = hN = hnN \; \checkmark$. The kernel of ϕ is,

$$Ker(\phi) = \{h \in H \mid \phi(h) = N\} \qquad (5.35)$$
$$= \{h \in H \mid hN = N\}$$
$$= \{h \in H \mid h \in N\}$$
$$= H \cap N.$$

According to the first isomorphism theorem,

$$H/Ker(\phi) \simeq HN/N. \qquad (5.36)$$

And therefore,

$$H/H \cap N \simeq HN/N. \qquad (5.37)$$

This concludes the proof of the second isomorphism theorem. $\boxed{\text{QED}\checkmark}$

5.3 Third Isomorphism Theorem

Theorem 5.7. *Let $N \subseteq H \subseteq G$ be groups such that $N \lhd G$ and $H \lhd G$. Then,*

1. $H/N \lhd G/N$

2. $(G/N)/(H/N) \simeq G/H$

3. *Every subgroup of G/N is of the form S/N for some subgroup S of G such that $N \subseteq S \subseteq G$*

4. *Every normal subgroup of G/N is of the form S/N for some normal subgroup S of G such that $N \subseteq S \subseteq G$.*

Proof of Part 1. To show that if $N \subseteq H \subseteq G$ are groups such that $N \lhd G$ and $H \lhd G$, then $H/N \lhd G/N$. Consider arbitrary elements $hN \in H/N$ and $gN \in G/N$.

$$
\begin{aligned}
(gN)hN(gN)^{-1} &= \\
gNhNNg^{-1} &= \\
ghg^{-1}N &
\end{aligned}
\tag{5.38}
$$

The commutation of N in the above equation follows because $N \lhd G$. Also, because $H \lhd G$ we have that $ghg^{-1} \in H$. Therefore $ghg^{-1}N \in H/N$, hence

$$(gN)hN(gN)^{-1} \in H/N. \tag{5.39}$$

Therefore,

$$H/N \lhd G/N \tag{5.40}$$

which is what we sought to prove. $\boxed{\text{QED}\checkmark}$

Proof of Part 2. To show that if $N \subseteq H \subseteq G$ are groups such that $N \lhd G$ and $H \lhd G$, then $(G/N)/(H/N) \simeq G/H$. Consider the map

$$\phi : G/N \to G/H \tag{5.41}$$

given by,

$$\phi(gN) = gH. \tag{5.42}$$

We claim that ϕ is a homomorphism. To see this, observe that for $i, j \in G$,

$$\phi(iNjN) = \tag{5.43}$$
$$\phi(ijN) =$$
$$ijH =$$
$$iHjH =$$
$$\phi(iN)\phi(jN). \; \checkmark$$

We claim that ϕ is a surjection. To see this, consider any $gH \in G/H$. And note that

$$\phi(gN) = gH. \tag{5.44}$$

The kernel of ϕ is given by,

$$Ker(\phi) = \{gN \in G/N \mid \phi(gN) = H\} \tag{5.45}$$
$$= \{gN \in G/N \mid gH = H\}$$
$$= \{gN \mid g \in H\}$$
$$= H/N.$$

Therefore by the first isomorphism theorem it follows that

$$(G/N)/(H/N) \simeq G/H \tag{5.46}$$

which is what we sought to prove. $\boxed{\text{QED}\checkmark}$

Proof of 3.: To prove that: Every subgroup of G/N is of the form S/N for some subgroup S of G such that $N \subseteq S \subseteq G$. The proof follows from the correspondence theorem which we proved in the previous chapter. As a reminder, the correspondence theorem states that if K is a subgroup of the quotient group G/H, then there exists a subgroup J of G given by,

$$J = \{xh \in G \mid h \in H \text{ and } xH \in K\}. \tag{5.47}$$

Therefore for any subgroup K of G/N, there exists a subgroup S of G given by

$$S = \{xn \in G \mid n \in N \text{ and } xN \in K\}. \tag{5.48}$$

It is easy to verify by showing mutual inclusion that

$$S/N = K. \tag{5.49}$$

It is also clear that $N \subseteq S \subseteq G$. Therefore we are done with the proof. $\boxed{\text{QED}\checkmark}$

Proof of 4.: To prove that: Every normal subgroup of G/N is of the form S/N for some normal subgroup S of G such that $N \subseteq S \subseteq G$. Again invoking our proof of the correspondence theorem, we know there exists some subgroup S of G given by

$$S = \{xn \in G | n \in N \text{ and } xN \in K\}. \tag{5.50}$$

All that remains for this proof is to show that if the subgroup S/N is normal in G/N then S is normal in G. Let $S/N \triangleleft G/N$, then

$$(gN)sN(gN)^{-1} \in S/N \tag{5.51}$$

for any $gN \in G/N$ and any $sN \in S/N$. It follows that,

$$(gN)sN(gN)^{-1} \in S/N \tag{5.52}$$
$$gsg^{-1}N \in S/N$$

In other words, for any $g \in G$ and $s \in S$, $gsg^{-1} \in S$. Therefore $S \triangleleft G$. $\boxed{\text{QED}\checkmark}$

5.4 Exercises

1. Consider the map

$$\sigma : \mathbb{Z}/15 \to \mathbb{Z}/3 \times \mathbb{Z}/5$$

   given by
$$\sigma(1) = (1, 2)$$

   a) Show that σ is a homomorphism.

   b) Show that σ is a bijection.

   c) List all the other admissible values of $\sigma(1)$ that yield an isomorphism.

2. Let $a, b \in \mathbb{N}$ such that a and b are relatively prime. Show that $\mathbb{Z}/ab \simeq \mathbb{Z}/a \times \mathbb{Z}/b$.

   a) If $\sigma : \mathbb{Z}/ab \to \mathbb{Z}/a \times \mathbb{Z}/b$, what general property must $\sigma(1)$ have to yield an isomorphism as desired?

   b) How many different admissible σ are there if $a = 7$ and $b = 11$?

   c) How many different admissible σ are there if $a = 10$ and $b = 3$?

3. Prove the first isomorphism theorem.

4. Let G and H be groups and G be simple. Let $\sigma : G \to H$ be an onto homomorphism.

   a) Show that $Ker(\sigma)$ is either trivial or all of G.

   b) If σ is non-trivial, show that $G \simeq H$.

5. Show that $\mathbb{Z}/5\mathbb{Z} \simeq \mathbb{Z}/5$.

6. Show that $\mathbb{R}^+/\mathbb{Z}^+ \simeq S^1$, where S^1 is the unit circle, i.e. $S^1 = \{z \in \mathbb{C}$ such that $\|z\| = 1\}$ under complex number multiplication.

7. State and prove the correspondence theorem.

8. State and prove the second isomorphism theorem.

9. State and prove the third isomorphism theorem.

10. Let S_n be the group of permutations of n letters, and A_n be the group of even permutations of n letters. Show that $S_n/A_n \simeq \mathbb{Z}/2$.

11. Show that $GL_n(\mathbb{R})/SL_n(\mathbb{R}) \simeq \mathbb{R}^\times$.

Chapter 6

The Dihedral Groups

6.1   Symmetries of the Regular Polygon

The dihedral group is the set of symmetries of the regular polygon under composition. The symbol D_{2n} denotes the dihedral group of order $2n$ and it represents the symmetries of the regular n-sided polygon. For instance, D_8 represents the symmetries of the square. Note that some texts use a different notation in which the symmetries of the regular n-gon are denoted D_n. The dihedral group of order 2n consists of rotations and of reflections. It is given by,

$$D_{2n} = \{e, r, r^2, \ldots r^{n-1}, \ c, cr, cr^2, \ldots cr^{n-1}\} \qquad (6.1)$$

where c represents a reflection about the horizontal and r represents a clockwise rotation by $2\pi/n$ degrees. And where,

$$c^2 = r^n = e. \qquad (6.2)$$

We can also check that

$$crc = r^{-1}. \qquad (6.3)$$

The above is called a relation. In the language of generators and relations, the dihedral group can be represented as,

$$D_{2n} = \langle c, r \rangle, \qquad (6.4)$$

where $crc = r^{-1}$, and $c^2 = r^n = e$.

6.2   Center of Dihedral Group

Since $crc = r^{-1}$, i.e. $cr = r^{-1}c$ it follows that the only rotation which commutes with the reflection is one for which $r = r^{-1}$. There can only exist such a rotation for even sided polygons, and the sole such rotation is the 180 degree rotation. Check to see that if you rotate say a square or a hexagon by 180 degrees, you get the exact same result in either direction i.e. clockwise or counterclockwise. Rotation by any other angle modulo 2π yields a different clockwise than counter-clockwise result. Therefore the center of the dihedral group is given by,

$$Z(D_{2n}) = \begin{cases} \{e, r^{n/2}\} & \text{if } n \text{ even} \\ \{e\} & \text{if } n \text{ odd} \end{cases} \tag{6.5}$$

6.3 Normal Subgroup of Dihedral Group

The group of rotations is the only non-trivial normal subgroup of the dihedral group. This is easily verified from the relation

$$crc^{-1} = crc = r^{-1}. \tag{6.6}$$

This also follows from the index two theorem, Theorem (3.8). Let N be the group of rotations. Then,

$$|N| = \frac{|D_{2n}|}{2}. \tag{6.7}$$

Therefore, by the index-2 thm,

$$N \triangleleft D_{2n}. \tag{6.8}$$

6.4   Exercises

1. Show that $crc = r^{-1}$ for $c, r \in D_6$.

2. Show that $crc = r^{-1}$ for $c, r \in D_{10}$.

3. What is the center of D_{10}?

4. What is the center of D_{12}?

5. Describe the center of D_{2n}.

6. Prove the index 2 theorem.

7. Show that the rotation subgroup is normal in D_{2n}.

8. Show that D_6 is isomorphic to S_3, the group of permutations of 3 letters.

9. List all elements of D_{10} and their respective orders.

Chapter 7

The Quaternion Group

7.1 Quaternion Group

The quaternion group, denoted Q_8, is the non-abelian group of order eight whose elements are $\pm i, \pm j, \pm k, \pm 1$ and whose multiplication relationship is given by,

$$i^2 = j^2 = k^2 = ijk = -1 \tag{7.1}$$
$$ij = k$$
$$jk = i$$
$$ki = j$$
$$ji = -k$$
$$ik = -j$$
$$kj = -i$$

The multiplication rule can be recalled by the following figure

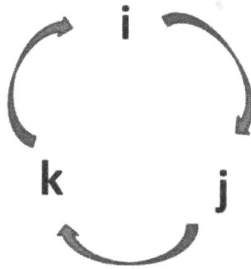

Figure 7.1: Quaternion multiplication

in which following the arrows clockwise yields the product with a positive sign, e.g. $ij = +k$, while going counterclockwise i.e. against the arrows yields the product with a negative sign e.g. $ji = -k$.

An alternate way of specifying the Quaternion group is in the language of generators and relations. In this parlance, the Quaternion group can be expressed as,

$$Q_8 = \langle i, j \mid i^4 = 1,\ i^2 = j^2,\ i^{-1}ji = j^{-1} \rangle \qquad (7.2)$$

7.2 Quaternion Group Matrix Representation

The matrix representation of the quaternion group is given by,

$$1 \rightarrow \begin{pmatrix} 1 & 0 \\ 0 & 1 \end{pmatrix} \qquad (7.3)$$

$$i \rightarrow \begin{pmatrix} \tilde{i} & 0 \\ 0 & -\tilde{i} \end{pmatrix} \qquad (7.4)$$

$$j \rightarrow \begin{pmatrix} 0 & 1 \\ -1 & 0 \end{pmatrix} \qquad (7.5)$$

$$k \rightarrow \begin{pmatrix} 0 & \tilde{i} \\ \tilde{i} & 0 \end{pmatrix} \qquad (7.6)$$

where i, j, and k are elements of Q_8 and $\tilde{i}$ is the complex number whose square is -1. To check that this indeed represents the quaternion group, one can check the multiplication relations as follows:

$$ij = \begin{pmatrix} \tilde{i} & 0 \\ 0 & -\tilde{i} \end{pmatrix} \begin{pmatrix} 0 & 1 \\ -1 & 0 \end{pmatrix} = \begin{pmatrix} 0 & \tilde{i} \\ \tilde{i} & 0 \end{pmatrix} = k \qquad (7.7)$$

$$jk = \begin{pmatrix} 0 & 1 \\ -1 & 0 \end{pmatrix} \begin{pmatrix} 0 & \tilde{i} \\ \tilde{i} & 0 \end{pmatrix} = \begin{pmatrix} \tilde{i} & 0 \\ 0 & -\tilde{i} \end{pmatrix} = i \qquad (7.8)$$

$$ki = \begin{pmatrix} 0 & \tilde{i} \\ \tilde{i} & 0 \end{pmatrix} \begin{pmatrix} \tilde{i} & 0 \\ 0 & -\tilde{i} \end{pmatrix} = \begin{pmatrix} 0 & 1 \\ -1 & 0 \end{pmatrix} = j \qquad (7.9)$$

$$ji = \begin{pmatrix} 0 & 1 \\ -1 & 0 \end{pmatrix} \begin{pmatrix} \tilde{i} & 0 \\ 0 & -\tilde{i} \end{pmatrix} = \begin{pmatrix} 0 & -\tilde{i} \\ -\tilde{i} & 0 \end{pmatrix} = -k \qquad (7.10)$$

$$ik = \begin{pmatrix} \tilde{i} & 0 \\ 0 & -\tilde{i} \end{pmatrix} \begin{pmatrix} 0 & \tilde{i} \\ \tilde{i} & 0 \end{pmatrix} = \begin{pmatrix} 0 & -1 \\ 1 & 0 \end{pmatrix} = -j \qquad (7.11)$$

$$kj = \begin{pmatrix} 0 & \tilde{i} \\ \tilde{i} & 0 \end{pmatrix} \begin{pmatrix} 0 & 1 \\ -1 & 0 \end{pmatrix} = \begin{pmatrix} -\tilde{i} & 0 \\ 0 & \tilde{i} \end{pmatrix} = -i \qquad (7.12)$$

7.3   Generalized Quaternion Group

The generalized quaternion groups are of the form,

$$Q_{4n} = \langle i, j \mid j^{2n} = 1,\ i^2 = j^n,\ i^{-1}ji = j^{-1} \rangle \qquad (7.13)$$

We can immediately see that in this language of generators and relations, Q_8 is indeed a special case of Q_{4n}. And in particular it is the case $n = 2$.

7.4 Exercises

1. Show that the quaternion group Q_8 is Hamiltonian, i.e. a non-abelian group in which every subgroup is normal.

2. Show that the quotient group $Q_8/\{\pm 1\}$ is isomorphic to the Klein four group $\mathbb{Z}/2 \times \mathbb{Z}/2$.

3. Show that the quaternion group is generated by i and j such that $i^4 = 1$, $i^2 = j^2$, $iji^{-1} = j^{-1}$.

4. Show that the commutator of Q_8 is isomorphic to $\mathbb{Z}/2$.

5. Show that the center of Q_8 is isomorphic to $\mathbb{Z}/2$.

6. Show that the generalized quaternion group Q_{4n} is Hamiltonian if and only if $n = 2$.

7. Show that $|Q_{4n}| = 4n$.

Chapter 8

Direct and Semi-Direct Products

8.1 Direct Products

Definition 8.1. Given groups U and V, the direct product of U and V is a group denoted by $U \times V$. It has elements of the form (u, v) where $u \in U$ and $v \in V$. The multiplication is defined by,

$$(u_1, v_1) \cdot (u_2, v_2) = (u_1 u_2, v_1 v_2). \tag{8.1}$$

The inverse is given by,

$$(u, v)^{-1} = (u^{-1}, v^{-1}). \tag{8.2}$$

And the identity is given by (e_U, e_V).

The definition extends naturally to higher dimensions. Given groups $U, V, H, K \ldots$, the group $U \times V \times H \times K \times \ldots$ has a multiplication operation defined by,

$$(u_1, v_1, h_1, k_1, \ldots) \cdot (u_2, v_2, h_2, k_2 \ldots) = \tag{8.3}$$
$$(u_1 u_2, v_1 v_2, h_1 h_2, k_1 k_2 \ldots).$$

The inverse is given by,

$$(u, v, h, k, \ldots)^{-1} = (u^{-1}, v^{-1}, h^{-1}, k^{-1}, \ldots). \tag{8.4}$$

And the identity is given by $(e_U, e_V, e_H, e_K, \ldots)$.

Theorem 8.2. *Let X, Y be groups, then $X \times Y$ is a group.*

Proof: We must show closure under multiplication, associativity, closure under inverse, and existence of identity. Closure under multiplication follows directly from the definition of the group multiplication. For associativity we have,

$$(x_1, y_1)[(x_2, y_2)(x_3, y_3)] = \quad\quad (8.5)$$
$$(x_1, y_1)(x_2 x_3, y_2 y_3) =$$
$$(x_1 x_2 x_3, y_1 y_2 y_3) =$$
$$(x_1 x_2, y_1 y_2)(x_3, y_3) =$$
$$[(x_1, y_1)(x_2, y_2)](x_3, y_3).$$

Finally, the inverse and the identity follow directly from the definition. $\boxed{\text{QED}\checkmark}$

A natural question to ask next is this: When does G equal the direct product of two of its given subgroups? This question motivates and introduces the concept of the *internal direct product*. The term *internal* describes the notion that the groups whose direct product are taken come from *inside* of G, i.e. they are subgroups of G. This distinction from the *external* direct product is subtle but important. The difference between the external and internal direct product is solely the way the problem is structured. i.e. what are the givens and what are the unknowns? For the internal direct product, the particular phrasing of the problem is this: Given G, do there exist subgroups $U, V \subset G$ such that $G = U \times V$? If so, what are they? The next theorem is about the conditions such subgroups U, V must satisfy.

Theorem 8.3. *Let $U, V \subset G$ be groups. Then $G = U \times V$ if and only if:*

1. $G = UV$

2. $U \cap V = \{e\}$

3. $U \lhd G$, *and* $V \lhd G$

Proof: The idea here is that the above features make the set $\{U, V\}$ a "basis" for the space G. We first prove that if the conditions 1-3 above hold, then $G = U \times V$. Assume conditions 1-3 hold. Then since $U \lhd G$, it follows that,

$$G/U = \{vU \mid v \in V\} \tag{8.6}$$

is a group of order $|V|$. In other words, quotienting by U partitions G into cosets such that each coset is of the form vU for some v in V. The vU are distinct from each other because $U \cap V = \{e\}$. Therefore,

$$G = \{uv \mid u \in U, \ v \in V\}. \tag{8.7}$$

Lemma 8.4. *Let $U, V \subset G$ be groups such that $U \lhd G$, $V \lhd G$, and $U \cap V = \{e\}$. Then $uv = vu$ for any $u \in U$ and $v \in V$.*

Proof of Lemma: Since $U \lhd G$ and $V \lhd G$, it follows that for any $u \in U$ and $v \in V$

$$v^{-1}uv \in U \quad \text{and} \quad uvu^{-1} \in V. \tag{8.8}$$

Therefore by closure under group multiplication,

$$(v^{-1}uv)u^{-1} \in U \tag{8.9}$$

and

$$v^{-1}(uvu^{-1}) \in V. \tag{8.10}$$

Therefore,

$$v^{-1}uvu^{-1} \in U \cap V = \{e\}. \tag{8.11}$$

Therefore,

$$uv = vu. \tag{8.12}$$

QED✓

Consider $g_1, g_2 \in G$ such that $g_1 = u_1 v_1$ and $g_2 = u_2 v_2$. Then

$$g_1 g_2 = (u_1 v_1)(u_2 v_2) = (u_1 u_2)(v_1 v_2). \qquad (8.13)$$

The above commutation is legal by Lemma (8.4). Altogether, we have shown that all elements in G are of the form uv where $u \in U$ and $v \in V$, and that the product of any two elements in G preserves this uv form. In other words, it is closed under multiplication, and satisfies the definition of the direct product group $U \times V$ via the isomorphisms:

$$uv \rightarrow (u, v) \qquad (8.14)$$
$$(u_1 v_1)(u_2 v_2) \rightarrow (u_1 u_2, v_1 v_2)$$

This concludes the "if" part of the proof. $\boxed{\text{QED}\checkmark}$.

Next we proceed with the "only if" part, i.e. we seek to show that if $G \simeq U \times V$ then conditions 1-3 above hold. Therefore let $G \simeq U \times V$. Then for all $g \in G$ it holds that $g = (u, v)$ where $u \in U$ and $v \in V$. That $G = UV$ follows naturally from the isomorphism $(u, v) \rightarrow uv$. To show that $U \cap V = \{e\}$, assume otherwise. i.e. that there exists some nontrivial $t \in U \cap V$. Then it follows that $t = t \cdot e \rightarrow (t, e)$. But also, $t = e \cdot t \rightarrow (e, t)$. Hence the mapping between G and $U \times V$ is not one-to-one, therefore $G \not\simeq U \times V$ which is a contradiction. Finally, to show normalcy of U and V in G, consider $vuv^{-1} \in G$ for some $u \in U$ and $v \in V$. This element maps to $(e, v)(u, v^{-1}) = (u, e)$ which maps from u. Hence we have not only shown that $U \triangleleft G$, but that a true and equivalent fact is $uv = vu$. i.e. the elements of U commute with the elements of V. The same argument shows that $V \triangleleft G$.

8.2 Semi-Direct Products

One key requirement for the internal direct product above was the normality of both subgroups in G. This guaranteed commutativity of elements between the subgroups, thereby allowing for closure under group multiplication. In the case of the internal semi-direct product, normality is only required of one of the subgroups. If both subgroups are normal in G, then that yields the direct product. Therefore, the semi-direct product can be thought of as a generalization of the direct product. Or alternatively, the direct product can be thought of as a special case of the semi-direct product. The internal semi-direct product is defined as follows:

Definition 8.5. Let $H, K \subset G$ be groups such that:

1. $HK = G$,

2. $H \cap K = \{e\}$, and

3. $K \lhd G$.

Then G is isomorphic to the semi-direct product of H and K. We write $G \simeq H \ltimes_\pi K$; where the group multiplication is given by:

$$(h_1, k_1)(h_2, k_2) = (h_1 h_2, \pi(h_2^{-1})k_1, k_2), \qquad (8.15)$$

where in the standard internal direct product π is simply conjugation, i.e.,

$$\pi(h_2^{-1})k_1 = h_2^{-1} k_1 h_2 \in K. \qquad (8.16)$$

The above membership in K is guaranteed by and follows from the definition of $K \lhd G$. The identity element is given by, (e_H, e_K). In the case of the internal semi-direct product $e_H = e_K = e_G$, hence in that case we simply denote the identity by (e, e). We can derive the inverse (h_2, k_2) of an arbitrary element (h_1, k_1) of $H \ltimes_\pi K$, by solving the equation,

$$(h_1, k_1)(h_2, k_2) = (e, e). \tag{8.17}$$

We get,

$$(h_1, k_1)(h_2, k_2) = (h_1 h_2, \pi(h_2^{-1})(k_1)k_2) \tag{8.18}$$

Therefore,

$$h_1 h_2 = e \tag{8.19}$$
$$h_2 = h_1^{-1}$$

and

$$\pi(h_2^{-1})(k_1)k_2 = e \tag{8.20}$$
$$k_2 = [\pi(h_2^{-1})(k_1)]^{-1} = \pi(h_2^{-1})k_1^{-1}.$$

Hence,

$$(h_1, k_1)^{-1} = (h_1^{-1}, \pi(h_2^{-1})k_1^{-1}). \tag{8.21}$$

And since $h_2^{-1} = h_1$, it follows that

$$(h_1, k_1)^{-1} = (h_1^{-1}, \pi(h_1)k_1^{-1}). \tag{8.22}$$

Theorem 8.6. *Let $H, K \subset G$ be groups such that $K \lhd G$, $H \cap K = \{e\}$, and $G = HK$. And let $\pi(h)$ be conjugation by h. Then $H \ltimes_\pi K$ is a group.*

Proof: Closure under multiplication, inverses, and the existence of the identity all follow directly from the definition. All that remains to prove is associativity. Consider

$$(h_1, k_1)[(h_2, k_2)(h_3, k_3)] = \qquad (8.23)$$
$$(h_1, k_1)(h_2 h_3, \pi(h_3^{-1})(k_2)k_3) =$$
$$\left(h_1 h_2 h_3, \pi(h_3^{-1} h_2^{-1})(k_1)\pi(h_3^{-1})(k_2)k_3\right)$$

and consider

$$[(h_1, k_1)(h_2, k_2)](h_3, k_3) = \qquad (8.24)$$
$$(h_1 h_2, \pi(h_2^{-1})(k_1)k_2)(h_3, k_3) =$$
$$\left(h_1 h_2 h_3, \pi(h_3^{-1})(\pi(h_2^{-1})(k_1)k_2)k_3\right) =$$
$$\left(h_1 h_2 h_3, \pi(h_3^{-1} h_2^{-1})(k_1)\pi(h_3^{-1})(k_2)k_3\right).$$

Therefore,

$$(h_1, k_1)[(h_2, k_2)(h_3, k_3)] = [(h_1, k_1)(h_2, k_2)](h_3, k_3) \quad (8.25)$$

$\boxed{\text{QED}\checkmark}$

External Semi-Direct Product

Above, we have alluded to the modularity of the semi-direct product by virtue of the homomorphism π. In the standard semi-direct product, we took π to be conjugation by h. In a more general frame work, we can describe the *external* semi-direct product between two — in our minds previously unrelated — groups S and T. We write

$$S \ltimes_\phi T \qquad (8.26)$$

as the external semi-direct product of S and T, where ϕ is a homomorphism defined by,

$$\phi : S \rightarrow Aut(T). \qquad (8.27)$$

The group multiplication is defined by,

$$(s_1, t_1)(s_2, t_2) = (s_1 s_2, \phi(s_2)\{t_1\}t_2) \qquad (8.28)$$

The identity is (e_S, e_T) and the inverse is $(s_1^{-1}, \pi(s_1)t_1^{-1})$.

8.3 Exercises

1. Let $U, V \subset G$ be groups such that $UV = G$, $U \cap V = \{e\}$, $U \triangleleft G$, and $V \triangleleft G$. Show that $G = U \times V$.

2. Let $U, V \subset G$ be groups such that $G = U \times V$. Show that $UV = G$, $U \cap V = \{e\}$, $U \triangleleft G$, and $V \triangleleft G$.

3. Let $\rho_K : H \times K \to K$ and $\rho_H : H \times K \to H$ be defined by $\rho_K((h, k)) := k$ and $\rho_H((h, k)) := h$.

   a) Show that ρ_H and ρ_K are homomorphisms.

   b) What are the kernels of ρ_H and ρ_K respectively?

4. Show that $\mathbb{Z}/12 \simeq \mathbb{Z}/3 \times \mathbb{Z}/4$.

5. Show that if m and n are relatively prime, then $\mathbb{Z}/mn \simeq \mathbb{Z}/m \times \mathbb{Z}/n$.

6. Show that the group $(\mathbb{C}, \cdot) \simeq \xi_\theta \times \mathbb{R}^+$, where ξ_θ is the unit circle $z = e^{i\theta}$, $\mathbb{R}^+$ are the positive real numbers, and $(\mathbb{C}, \cdot)$ are the complex numbers under complex number multiplication.

7. Show that D_{12}, the dihedral group of order 12, is isomorphic to $\langle r^2, c \rangle \times \{e, r^3\}$; where $r^6 = c^2 = e$.

8. Show that D_{20}, the dihedral group of order 20, is isomorphic to $\langle r^2, c \rangle \times \{e, r^5\}$; where $r^{10} = c^2 = e$.

9. Let D_{4n} denote the dihedral group of order $4n$, and let n be odd. Show that $D_{4n} \simeq \langle r^2, c \rangle \times \{e, r^n\}$; where $r^{2n} = c^2 = e$.

10. Show that if the homomorphism ϕ is non-trivial, then $H \ltimes_\phi K$ is not abelian.

11. Show that the quaternion group is a splitting-simple group, i.e. it cannot be expressed as a semi-direct product of its subgroups.

12. Show that $D_{2n} \simeq \mathbb{Z}/2 \ltimes \mathbb{Z}/n$.

Chapter 9

Symmetric and Alternating Groups

9.1 Symmetric Group

The symmetric group on n letters, denoted S_n is the group of permutations of n letters. It forms a group under composition. The order of the group is n! And this derives because there are n choices for the first letter, and (n-1) choices for the 2nd letter,... and 1 choice for the nth letter.

$$|S_n| = n! \qquad (9.1)$$

Example 9.1. S_3 is the group of permutations of 3 letters. By our formula it has $3! = 6$ elements. That means given three letters a_1, a_2, a_3 in that order, there are six different ways to order the elements:

$$
\begin{array}{ll}
1) & a_1, a_2, a_3 \qquad\qquad (9.2) \\
2) & a_1, a_3, a_2 \\
3) & a_3, a_1, a_2 \\
4) & a_3, a_2, a_1 \\
5) & a_2, a_1, a_3 \\
6) & a_2, a_3, a_1
\end{array}
$$

The elements of S_3 consist of the permutation operations that yield the different permutations above. The corresponding elements of S_n can be represented as:

$$
\begin{array}{ll}
1) & e \qquad\qquad (9.3) \\
2) & (23) \\
3) & (123) \\
4) & (13) \\
5) & (12) \\
6) & (132)
\end{array}
$$

The above notation is called *cycle notation* and is one of a number of ways to represent elements of the symmetric group. e is the identity and it means "leave all elements unpermuted". (23) is called a 2-cycle and it means "move the letter in position 2 to position 3, and the letter in position 3 to position 2". (123) is called a 3-cycle and it means "move the letter in position 1 to position 2, move the letter in position 2 to position 3, and move the letter in position 3 to position 1. There are other ways to represent elements of S_n. For instance, (123) can also be represented as,

$$(123) = \begin{pmatrix} a_1 & a_2 & a_3 \\ a_3 & a_1 & a_2 \end{pmatrix} \tag{9.4}$$

9.2 Multiplication in Symmetric Group

In cycle notation, we apply the permutations from right to left as we do for compositions of functions. The order of multiplication does not matter for disjoint cycles, but does for non-disjoint cycles. For instance, one can verify that,

$$\begin{pmatrix} a_1 & a_2 & a_3 \\ a_3 & a_1 & a_2 \end{pmatrix} \begin{pmatrix} a_1 & a_2 & a_3 \\ a_1 & a_3 & a_2 \end{pmatrix} = \tag{9.5}$$
$$\begin{pmatrix} a_1 & a_2 & a_3 \\ a_3 & a_2 & a_1 \end{pmatrix}.$$

In cycle notation, this says,

$$(132)(23) = (13). \tag{9.6}$$

The left hand side of the above equation does not commute, since

$$(23)(132) = (12) \neq (13). \tag{9.7}$$

On the other hand,

$$(13)(24) = (24)(13). \tag{9.8}$$

9.3 Alternating Groups

The Alternating group on n letters, denoted A_n, is the subgroup of S_n consisting of even permutations. Examples of even permutations include disjoint 2-cycles such as $(13)(24)$ or 3-cycles such as (123). Note that cycles of odd length are even permutations while cycles of even length are odd permutations.

$$|A_n| = \frac{n!}{2} \tag{9.9}$$

Theorem 9.2. $A_n \lhd S_n$

Proof: Since $|A_n| = |S_n|/2$, it follows from the index 2 theorem that $A_n \lhd S_n$.

9.4 Exercises

1. Show that S_n/A_n is isomorphic to $\mathbb{Z}/2$.

2. Show that $A_n \lhd S_n$.

3. Show that the Klein-four group $\mathbb{Z}/2 \times \mathbb{Z}/2$ is a proper normal subgroup of the Alternating group A_4.

4. Show that any 5-cycle can be decomposed into a product of four 2-cycles.

5. Show that any n-cycles where n is odd, can be decomposed into a product of n-1 2-cycles.

6. Show that S_3 is isomorphic to D_6.

7. Show that the 3-cycles generate A_n.

Chapter 10

Group Actions

Actions speak louder than words but not nearly as often.

–Mark Twain

Definition 10.1. Given a group G and a set X, a *group action* of G on X is a homomorphism π between G and the group of bijections from X to itself.

$$\pi : G \to Bij(X, X) \tag{10.1}$$

Therefore by homomorphism, given any $g, h \in G$ and $x, y, z \in X$, if $\pi(g)y = z$ and $\pi(h)x = y$ then $\pi(gh) = \pi(g) \circ \pi(h)x = \pi(g)y = z$. Note that the set X can be any arbitrary collection. In particular, it can be a group. It can even be the group G itself, or a collection of subgroups of G, or a collection of pebbles of various colors. The term *G-action on X* can be used to indicate group action of a group G acting on a set X.

10.1 Left Regular Action

One very important example of a group action of a group G acting on itself is called the left regular action. It is defined simply as left multiplication as follows:

$$\pi : G \to Bij(G, G) \tag{10.2}$$

such that,

$$\pi(g)h = gh. \tag{10.3}$$

We can easily check that this is a homomorphism. i.e. that $\pi(gh) = \pi(g) \circ \pi(h)$. Given some $g, h, x \in G$, $\pi(gh)x = ghx = \pi(g)\pi(h)x = \pi(g)hx = ghx$. We will see in the following chapter that the left regular action yields us a powerful theorem called Cayley's theorem. It states that every finite group of order n is isomorphic to a subgroup of the symmetric group on n letters.

10.2   Orbits and Stabilizers

Definition 10.2. Given a group action of a group G acting on a set X. And let x be an element in X. Then the *orbit* of x is defined as the set of all elements in X to which x is sent by some element of G.

The orbit of x under the G-action π is denoted Ω_x and is given by,

$$\Omega_x = \{\pi(g)x \mid g \in G\}. \tag{10.4}$$

Definition 10.3. Given a G-action π, the *stabilizer* of x in G, denoted $Stab_G(x)$, is the set of elements of $g \in G$ for which $\pi(g)x = x$.

i.e.,

$$Stab_G(x) = \{g \in G \mid \pi(g)x = x\}. \tag{10.5}$$

Theorem 10.4. $Stab_G(x)$ *is a subgroup of* G.

Proof: Since all elements of $Stab_G(x)$ are elements of G, it follows that $Stab_G(x)$ is a subset of G. All that remains is to show that $Stab_G(x)$ is itself a group. We will show the group properties in turn. $Stab_G(x)$ contains the identity because group actions are homomorphisms by definition, and homomorphisms map the identity of the domain group to the identity of the image group, in this case $Bij(X, X)$. For closure under multiplication, consider $g \in Stab_G(x)$ and $h \in Stab_G(x)$. Then $\pi(h)x = \pi(g)x = x$. Therefore $\pi(h)\pi(g)x = \pi(h)x = x$. For closure under inverses, consider $g \in Stab_G(x)$. Then $\pi(g)x = x$. Multiplying both sides by $\pi^{-1}(g)$ yields $\pi^{-1}(g)\pi(g)x = \pi^{-1}(g)x$. Since π is a homomorphism, the left hand side of the prior equation is equal to $\pi(g^{-1})\pi(g)x = \pi(g^{-1}g)x = \pi(e)x = x$. Therefore, $\pi(g^{-1})x = x$, hence $g^{-1} \in Stab_G(x)$. Finally, associativity follows directly from homomorphism. If $g, h, l \in Stab_G(x)$, then $\pi(gh)\pi(l) = \pi(g)\pi(hl)x = x$.

10.3 Transitive Group Actions

Definition 10.5. A group action $\pi : G \to Bij(X, X)$ is called *transitive* if given any $x \in X$ and any $y \in X$, there exists a $g \in G$ such that $\pi(g)x = y$.

Theorem 10.6. *Let π be a G-action on a set X. And let $x, y \in X$ such that $\pi(g)x = y$. It follows that $Stab_G(y) = gStab_G(x)g^{-1}$.*

Proof: Consider an arbitrary element of $gStab_G(x)g^{-1}$, i.e. gsg^{-1} where $s \in Stab_G(x)$. Then,

$$\pi(gsg^{-1})y = \pi(g)\pi(s)\pi(g^{-1})y \qquad (10.6)$$

Then since

$$\pi^{-1}(g)y = \pi(g^{-1})y = x, \qquad (10.7)$$

it follows that,

$$\begin{aligned}
\pi(gsg^{-1})y &= \pi(g)\pi(s)\pi(g^{-1})y = \qquad (10.8) \\
\pi(gsg^{-1})y &= \pi(g)\pi(s)x = \\
\pi(gsg^{-1})y &= \pi(g)x = \\
\pi(gsg^{-1})y &= y.
\end{aligned}$$

Therefore,

$$gStab_G(x)g^{-1} \subset Stab_G(y). \qquad (10.9)$$

And since $|gStab_G(x)g^{-1}| = |Stab_G(y)|$, it follows that,

$$Stab_G(y) = gStab_G(x)g^{-1}. \qquad (10.10)$$

$\boxed{\text{QED}\checkmark}$

10.4   Orbit-Stabilizer Theorem

Theorem 10.7. *Given a G-action on a set X, and $x \in X$. The order of the orbit of x, Ω_x, is given by,*

$$\Omega_x = \frac{|G|}{|Stab_G(x)|}. \tag{10.11}$$

Proof: Consider a map ξ between coset space $G/Stab_G(x)$ and Ω_x, defined by,

$$\xi : G/Stab_G(x) \to \Omega_x \tag{10.12}$$

where

$$\xi(gStab_G(x)) = \phi(g)x. \tag{10.13}$$

If ξ is a bijection, then we are done with the proof. This is because $|G/Stab_G(x)| = \frac{|G|}{|Stab_G(x)|}$, hence bijection implies $|\Omega_x| = |G/Stab_G(x)| = \frac{|G|}{|Stab_G(x)|}$ also. Note that the map ξ is *onto*, because by definition, given any $x' \in \Omega_x$ there exists a $g' \in G$ such that $\pi(g')x = x'$. Hence, $\xi(g'Stab_G(x)) = x'$. All that remains is to show ξ is one-to-one. Consider some $g, g' \in G$ such that $\xi(gStab_G(x)) = \xi(g'Stab_G(x))$. Then $\phi(g)x = \phi(g')x$. Therefore $\phi(g'^{-1}g)x = x$. Hence, $g'^{-1}g \in Stab_G(x)$, and therefore $g \in g'Stab_G(x)$. It follows that if $\xi(gStab_G(x)) = \xi(g'Stab_G(x))$ then $gStab_G(x) = g'Stab_G(x)$. i.e. ξ is one-to-one. This concludes the proof. $\boxed{\text{QED}\checkmark}$

10.5   Normalizers

Definition 10.8. Let G be a group and H a subgroup of G. Then the *normalizer* of H, denoted N(H), is given by,

$$N(H) = \{x \in G \mid xHx^{-1} = H\} \tag{10.14}$$

In other words, the normalizer of H is the set of elements in G for which H behaves as though it is normal. If H is normal in G then $N(H) = G$.

10.6   Exercises

1. Given a homomorphism σ, show that the identity is necessarily an element of the kernel of σ.

2. Given a group G with subgroup H for which $[G : H] = p$, consider the homomorphism $\pi : G \rightarrow S_p$, which arises via left regular action of G on $[G : H]$, where S_p is the symmetric group on p letters. Show that $\pi(G)$ is a subgroup of S_n.

3. Given a group G with subgroup H for which $[G : H] = k$ and the homomorphism $\pi : G \rightarrow S_k$, show that the kernel of π is the largest normal subgroup of G contained in H.

4. Let G be a group with a subgroup H such that $[G : H] = p$ where p is the smallest prime number divisor of G. Show that H is normal in G.

5. Given a group G acting on a set X. Show that the stabilizer of x in G is a subgroup of G.

6. Prove the orbit-stabilizer theorem.

7. Consider D_{12}, the dihedral group of order 12 i.e. the symmetries of a regular hexagon. And let X be the set of edges, Y the set of vertices, and Z the set of diagonals of a regular hexagon. Consider the action of D_{12} on X, on Y, and on Z respectively.

   a) What is the stabilizer of a vertex?

   b) What is the stabilizer of an edge?

   c) What is the stabilizer of a diagonal?

8. Let G be the $GL_n(\mathbb{R})$ and X be $\mathbb{R}^n$. What is the stabilizer of the standard basis vector e_j?

9. Let H be a subgroup of a group G. Show that the normalizer of H is a subgroup of G.

10. Consider the groups $S \subset H \subset G$. Show that $S \triangleleft H$ if and only if $H \subset N(S)$.

Chapter 11

Cayley's Theorem

Arthur Cayley (1821-1895) *Arthur Cayley was a British*

mathematician born in Richmond England in 1821. He graduated Cambridge as First Wrangler. He worked as a lawyer, while doing Mathematics and publishing prolifically. At the age of 42 funds were made available to establish the Sadleirian Chair of Mathematics at Cambridge. Cayley was offered the position and immediately quit his successful law practice. He expressed no regrets at his shrunken income, since the professorship freed him to focus on his mathematical pursuits. We have Cayley to thank for the modern definition of a group. Prior to this introduction, the term 'group' meant permutation group. Both Cayley and Camille Jordan contributed significantly to what we now refer to as "Cayley's theorem," the topic of this chapter. On the invitation of his friend J.J. Sylvester, Cayley spent most of 1882 in Baltimore Maryland, lecturing on Abelian and Theta functions at Johns Hopkins University.

11.1 Cayley's Theorem

Theorem 11.1. *Cayley's theorem[7,8] states that every finite group G is isomorphic to a subgroup of $S_{|G|}$.*

First let us examine an example of what Cayley's theorem says. Consider the group $\mathbb{Z}/3$. Below we show an isomorphism between $\mathbb{Z}/3$ and a subgroup of S_3.

Table 11.1: $\mathbb{Z}/3$ Example of Cayley's Theorem

	0	1	2
$0 \simeq \sigma_e \rightarrow$	0	1	2
$1 \simeq \sigma_1 \rightarrow$	1	2	0
$2 \simeq \sigma_2 \rightarrow$	2	0	1

We see that in the above table, each element of $\mathbb{Z}/3$ is associated with a permutation of the elements of $\mathbb{Z}/3$, and therefore with an element of S_3. One can easily check that $\{\sigma_e, \sigma_1, \sigma_2\}$ form a group and hence a subgroup of S_3. For instance, $\sigma_e \circ \sigma_1 = \sigma_1$, just as $0 + 1 = 1$ in $\mathbb{Z}/3$. Similarly, $\sigma_e \circ \sigma_2 = \sigma_2$, just as $0 + 2 = 2$ in $\mathbb{Z}/3$. A bit more interestingly, one can examine the table to verify that $\sigma_1 \circ \sigma_2 = \sigma_e$, just as $1 + 2 = 0$ in $\mathbb{Z}/3$. What we just showed for $\mathbb{Z}/3$ is that the mapping of its elements to permutations as shown in the table is a homomorphism. Also each element is assigned to a different permutation, so it is one-to-one, i.e. injective. Therefore it is isomorphic to a subgroup of S_3. In particular, it is isomorphic to a proper subgroup of S_3 because $|S_3| = 3! = 6$, while $|\mathbb{Z}/3| = 3$. Note for instance, that the permutation that takes $\{0, 1, 2\}$ to $\{2, 1, 0\}$ is not represented in our $\mathbb{Z}/3$ table. Nor is the permutation that takes $\{0, 1, 2\}$ to $\{1, 0, 2\}$. This will be true in general, because $|S_n| = n! < |G| = n$.

The power and significance of Cayley's theorem lies in the fact that every finite group is isomorphic to some symmetric group. Therefore everything that is true of symmetric groups is true of groups in general. It follows that when one understands the symmetric group, one understands all groups.

This becomes even more striking when one considers that the symmetric group is itself isomorphic to the permutation matrices, a subgroup of the group of orthogonal matrices. Therefore, understanding those innocent looking square matrices with nothing but zeros and ones in their entries is equivalent to having a full understanding of finite group theory. This profound reality is yet to be developed and exploited in the study of algebra. This is possibly because historically, the concept and study of groups in the more abstract sense came after the notion of symmetric groups. For the same reason, the more abstract notion of groups is generally perceived as more "advanced" study than the humble study of the most benign-appearing of the matrices. Regardless of custom however, the truth is the truth, and here it says all of finite group theory can be embedded into the study of orthogonal matrices!

Now that we understand what Cayley's theorem says and why it is so important, let us proceed to prove it in general.

Proof of Cayley's Theorem: The proof of Cayley's theorem is simple. Consider the left regular action of a group G on itself. Then each element $g \in G$ yields a permutation of the elements of G. i.e. the left regular action induces a map ϕ from G to $Bij(G, G)$, defined by $\phi(g_j) = \phi_{g_j}$ where $\phi_{g_j}(g_1, g_2, \ldots, g_n) = (g_j g_1, g_j g_2, \ldots, g_j g_n)$. For nontrivial g, the permutation is non-trivial. This is true because $g \cdot e = g \neq e$. It follows therefore that the kernel of the map is trivial. Therefore if this map is a homomorphism, then because it has trivial kernel it is injective. All that remains is to show that this left regular action-induced map is indeed a homomorphism. This follows directly from the associativity of G. i.e. for $g, x, h \in G$, it is true that $gh \cdot x = g \cdot (h \cdot x) = ghx$. i.e. $\phi_{gh}(x) = ghx = \phi_g(\phi_h(x)) = \phi_g(hx) = ghx$. Therefore ϕ is an isomorphism between G and $\phi(G) \subsetneq S_n$. $\boxed{\text{QED}\checkmark}$

11.2 Corollary to Cayley's Theorem

Corollary 11.2. *Let G be a finite group of order n, and H a subgroup of G. If $|G|$ does not divide $[G : H]!$, then H contains a non-trivial group $N \lhd G$.*

Proof of Corollary: If $|G|$ does not divide $|S_{[G:H]}| = [G : H]!$, then G is not isomorphic to any subgroups of $S_{[G:H]}$. This is because by Lagrange's theorem, the order of a subgroup must divide the order of the group. Now consider the homomorphic map given by,

$$\Phi : G \to Bij(G/H, G/H) \tag{11.1}$$

given by,

$$\Phi(g) = \phi_g, \tag{11.2}$$

where, $\phi_g : G/H \to G/H$ is given by,

$$\phi_g(xH) = gxH. \tag{11.3}$$

The kernel of Φ is the largest normal subgroup of G contained in H –verify this as an exercise. Since Φ is a homomorphism, it follows that if Φ is not be an isomorphism, it must be the case that its kernel is non-trivial. Therefore, $Ker(\Phi) \neq e$. Hence $e \neq N = Ker(\Phi) \lhd G$. $\boxed{\text{QED}\checkmark}$

11.3 Symmetric Group Revisited: Historical Context

As we discussed above, Cayley's theorem draws the symmetric group back into center stage as more fundamental than our more abstract notion of groups in general. Its fundamental nature undoubtedly played a role in the primality of order in which it was discovered. It has been explicitly apparent to all cultures since antiquity. In particular, it has long been familiar with some level of sophistication to ancient Babylonian, Egyptian, Ethiopian, Chinese, Timbuktian, and Indian mathematicians.[9] More recently, in the 12th century AD, in his treatise, the Lilavati[10], Indian mathematician Bhaskara II wrote in so many words that $|S_n| = n!$. The masterful work of Arthur Cayley some 600 years later was an extension of millennia of work spanning several cultures and civilizations. That work continues to this day. Such is the nature of Mathematics. The symmetric group continues to take its place in center stage through the remainder of this book, as well as all of Algebra. Here we look at one more of a plethora of theorems which arise from it. You may have encountered it as an exercise in the previous chapter.

Theorem 11.3. *If G is a group with a subgroup H such that $[G : H] = p$ where p is the smallest prime number divisor of G, then H is normal in G.*

Proof: Consider the homomorphism, $\pi : G \to S_p$, which arises via left regular action of G on $[G : H]$, where S_p is the symmetric group on p letters. Since $\pi(G)$ is a subgroup of S_p, it follows that $|\pi(G)| \mid p!$. From the first isomorphism theorem implies that $\pi(G) \simeq G/N$, where N is the kernel of π i.e. N is the largest normal subgroup of G contained in H. It follows that $|\pi(G)| = |G/N|$, therefore $|G/N| \mid p!$. Since $|N \subset H|$, $|G/N| \geq |G/H| = p$. Assessing the two facts together–i.e. $|G/N| \mid p!$ and $|G/N| \geq p$– we see that $|G/N| = p = |G/H|$. Therefore $N = H$, hence $H \lhd G$. $\boxed{\text{QED}\checkmark}$

11.4 Exercises

1. To what subgroup of S_5 is $\mathbb{Z}/5$ isomorphic?

2. List three permutations of 5 letters which are not elements of $\mathbb{Z}/5$.

3. To what subgroup(s) of S_8 is the dihedral group of order 8 isomorphic?

4. To what subgroup(s) of S_8 is $\mathbb{Z}/2 \times \mathbb{Z}/2 \times \mathbb{Z}/2$ isomorphic?

5. To what subgroup(s) of S_8 is $\mathbb{Z}/8$ isomorphic?

6. Prove Cayley's theorem.

7. Let G be a group with a subgroup H such that $[G : H] = p$ where p is the smallest prime number divisor of G. Prove that H is normal in G.

8. Consider the map ϕ induced by the left regular action of G acting on the coset space G/H. $\phi : G \to Bij(G/H, G/H)$. Show that the kernel of ϕ is the largest normal subgroup contained in H.

9. Let G be a simple group of order 60. Show that it cannot have a subgroup of order 15.

10. Can a simple group of order 168 have a subgroup of order 28? Why or why not?

11. Prove that a simple group of order 60 can have no subgroups of order 15. (Hint: Use the corollary to Cayley's theorem).

12. Prove that every symmetric group is isomorphic to a group of orthogonal matrices.

13. Consider a group action $\sigma : G \to Bij(X, X)$. It is said to be *faithful* if: "(for all $x \in X$ $\sigma(g)(x) = x$) $\Rightarrow$ $x = e$". Show that left multiplication is a faithful group action of any group on itself.

Chapter 12

Conjugacy and Class

Conjugate Cardinals

A pair of robins, I thought they were,
Way up high up in the atmosphere,
But as they perched from their orbit, I caught a closer stare,

The male red and bright, fed the female after flight,
Gosh, what a conjugate pair they were,
Cardinals, in matrimonial flair.

–S.G. Odaibo

12.1 Conjugation

One very special group action that must be embedded in the mind is that of conjugation. Like all group actions, it can act on a set or on a group. It is defined by,

$$G \cdot X = \{gxg^{-1} \mid g \in G , x \in X\} \qquad (12.1)$$

The orbits in this group action are called *conjugacy classes*. The conjugacy class of an element x is denoted C_x and defined,

$$C_x = \{gxg^{-1} \mid g \in G\}. \qquad (12.2)$$

We say gxg^{-1} is the conjugation of x by g. Under this special group action, the stabilizer of the element x is called the *centralizer* of x. It is denoted $C(x)$ and is defined,

$$C(x) = \{g \in G \mid gxg^{-1} = x\}. \qquad (12.3)$$

The term centralizer is suggestive. It is because from the perspective of x, $C(x)$ are the set of elements that act like the center, $Z(G)$. i.e. the set of elements that commute with x. Since if

$$gxg^{-1} = x, \qquad (12.4)$$

it follows that,

$$gx = xg. \qquad (12.5)$$

Since this is a group action, the orbit-stabilizer theorem naturally applies. It is given by,

$$|C_x| = \frac{|G|}{|C(x)|}. \qquad (12.6)$$

12.2 Conjugation as Equivalence Relation

Theorem 12.1. *Conjugacy classes are equivalence classes*

Proof: To prove this, we must show reflexivity, symmetry, and transitivity. To show reflexivity, i.e. $x \sim x$, we conjugate x by x. We get,

$$xxx^{-1} = x. \tag{12.7}$$

To show symmetry, i.e. $x \sim y \Rightarrow y \sim x$, note that $x \sim y$ implies that there exists some $g \in G$ such that,

$$gxg^{-1} = y. \tag{12.8}$$

Multiplying the right hand side by g and the left hand side by g^{-1}, we get,

$$x = gyg^{-1}. \tag{12.9}$$

Next we show transitivity, i.e. $x \sim y$ and $y \sim z$ then $x \sim z$. Let $x \sim y$ and $y \sim z$. Then there exists $h, g \in G$ such that,

$$gxg^{-1} = y \tag{12.10}$$

and

$$hyh^{-1} = z. \tag{12.11}$$

Substituting for y in the above equation, we get,

$$h(gxg^{-1})h^{-1} = z. \tag{12.12}$$

Rearranging yields,

$$(hg)x(hg)^{-1} = z. \tag{12.13}$$

Therefore,

$$x \sim z. \tag{12.14}$$

This concludes the proof. $\boxed{\text{QED}\checkmark}$

Corollary 12.2. *Conjugacy classes are either disjoint or identical.*

Proof: Since the conjugacy classes are equivalence classes, it follows directly that they are either disjoint of identical. $\boxed{\text{QED}\checkmark}$

Remark. The action of conjugation on a group of a set, partitions the group or set into a disjoint union of its conjugacy classes.

Theorem 12.3. *If G is an abelian group, then every conjugacy class is a singleton, i.e. contains only one element.*

Proof: If G is abelian, then for every $x, g \in G$, it follows that $gxg^{-1} = gg^{-1}x = x$. Therefore $C_x = \{x\}$.

12.3   The Class Equation

For the same reason that each element of an abelian group is the sole resident of its conjugacy class, the conjugacy classes of elements in the center of any group are singletons. In addition, we have seen that conjugacy partitions a group (or set) into a disjoint union of conjugacy classes. Putting these two facts together, we can write out an equation which classifies the group elements by conjugacy class. We get the *class equation,*

$$|G| = |Z(G)| + \sum_{x \ rep} |C_x|, \tag{12.15}$$

where $x \notin Z(G)$ and *'x rep'* indicates that the sum is taken over representative elements from each conjugacy class, i.e. the members of each conjugacy class are counted only once. If the group action is acting on a set X, then the analogous equation is,

$$|X| = |\zeta(G)| + \sum_{x \ rep} |C_x|, \qquad (12.16)$$

where $\zeta(G)$ is the set of fixed point in X under the group action, and $x \notin \zeta(G)$. The orbit-stabilizer theorem applies in each case, yielding us the equations,

$$|G| = |Z(G)| + \sum_{x \ rep} \frac{|G|}{|C(x)|} \qquad (12.17)$$

and

$$|X| = |\zeta(G)| + \sum_{x \ rep} \frac{|G|}{|C(x)|} \qquad (12.18)$$

respectively.

Example 12.4. Abelian groups have singleton conjugacy classes.

As we proved in Theorem (12.3), the conjugacy classes of elements of an abelian group are all singletons. For instance, each conjugacy class of each group of order less than 5 is simply a singleton. But more generally, this holds true for all abelian groups. The smallest non-abelian group is S_3, and is of order 6. It is the Dihedral group of order 6. More generally, we treat the dihedral group D_{2n} in the next section.

12.4 Conjugacy Classes of Dihedral Groups

Here we take a look at the conjugacy classes of the dihedral group. Recall that the dihedral group of order $2n$ is given by,

$$D_{2n} = \{r, r^2, \ldots, r^n, cr, cr^2, \ldots, cr^n\}, \qquad (12.19)$$

where

$$r^n = c^2 = e$$

and

$$crc = r^{-1}.$$

In a geometric interpretation the dihedral groups are the symmetry groups of regular polyhedra. The r^k are rotations and the cr^k are reflections of the regular polyhedra. Conjugating a rotation r^k by an arbitrary rotation r^t, we get,

$$r^t r^k r^{-t} = r^k. \tag{12.20}$$

Conjugating r^k by an arbitrary reflection cr^t, we get,

$$cr^t r^k (cr^t)^{-1} = cr^t r^k (r^{-t}c) \tag{12.21}$$
$$= r^{-t} r^{-k} r^t = r^{-k}.$$

It follows that the conjugacy class of a rotation is given by,

$$C_{r^k} = \{r^k, r^{-k}\}, \tag{12.22}$$

i.e. the conjugacy class consists of the rotation element and its inverse. Now conjugating a reflection element cr^k by an arbitrary rotation r^t, we get,

$$r^t cr^k r^{-t} = cr^{-t} r^k r^{-t} = cr^{k-2t}. \tag{12.23}$$

Conjugating cr^k by an arbitrary reflection cr^t, we get,

$$(cr^t)cr^k(cr^t)^{-1} = (cr^t)cr^k(r^{-t}c) \tag{12.24}$$
$$= cr^t r^{-k} r^t$$
$$= cr^{2t-k}.$$

It follows that,

$$C_{cr^k} = \{cr^{k-2t}, cr^{2t-k}\}. \tag{12.25}$$

In terms of a geometric interpretation, there are 2 cases to consider:

Case I: n is odd.

Case Ia: k *is odd.* Therefore $n - k$ is even. Therefore $r^{-k} \equiv r^{n-k}$ is of type r^{even}. And since *even* + *even* = *even*, it follows that r^{2t-k} is of type r^{even}. And r^{k-2t} is of type $r^{odd-even} = r^{odd}$. Therefore,

$$
\begin{aligned}
C_{cr^k} &= \{cr^{k-2t}, cr^{2t-k}\} \\
&= \{cr^{odd}, cr^{even}\} \\
&= \{cr, cr^2, \ldots, cr^n\}
\end{aligned}
\tag{12.26}
$$

i.e. in this case all the reflections form a single conjugacy class.

Case Ib: k *is even.* Therefore $n - k$ is odd. Therefore $r^{-k} \equiv r^{n-k}$ is of type r^{odd}. And since *even* + *odd* = *odd*, it follows that r^{2t-k} is of type r^{odd}. And r^{k-2t} is of type $r^{even-even} = r^{even}$. Therefore,

$$
\begin{aligned}
C_{cr^k} &= \{cr^{k-2t}, cr^{2t-k}\} \\
&= \{cr^{even}, cr^{odd}\} \\
&= \{cr, cr^2, \ldots, cr^n\}
\end{aligned}
\tag{12.27}
$$

i.e. in this case also, all the reflections form a single conjugacy class.

Case II: n is even.

Case IIa: k *is odd.* Therefore $n - k$ is odd. Therefore $r^{-k} \equiv r^{n-k}$ is of type r^{odd}. And since *even* + *odd* = *odd*, it follows that r^{2t-k} is of type r^{odd}. And r^{k-2t} is of type $r^{odd-even} = r^{odd}$. Therefore,

$$
\begin{aligned}
C_{cr^k} &= \{cr^{k-2t}, cr^{2t-k}\} \qquad (12.28)\\
&= \{cr^{odd}, cr^{odd}\}\\
&= \{cr^{odd}\}\\
&= \{cr, cr^3, \ldots, cr^{n-1}\}.
\end{aligned}
$$

Case IIb: *k is even.* Therefore $n - k$ is even. Therefore $r^{-k} \equiv r^{n-k}$ is of type r^{even}. And since $even + even = even$, it follows that r^{2t-k} is of type r^{even}. And r^{k-2t} is of type $r^{even-even} = r^{even}$. Therefore,

$$
\begin{aligned}
C_{cr^k} &= \{cr^{k-2t}, cr^{2t-k}\} \qquad (12.29)\\
&= \{cr^{even}, cr^{even}\}\\
&= \{cr^{even}\}\\
&= \{cr^2, cr^4, \ldots, cr^n\}.
\end{aligned}
$$

It follows that for n odd, the conjugacy classes are given by,

$$
\begin{aligned}
G = \{e\} \cup \{r, r^{n-1}\} \cup \{r^2, r^{n-2}\} \ldots \qquad (12.30)\\
\ldots \cup \{r^{(n-1)/2}, r^{(n+1)/2}\}\\
\cup \{cr, cr^2, \ldots, cr^n\}
\end{aligned}
$$

In this case where n is odd, there are a total of $1 + \frac{(n-1)}{2} + 1$ conjugacy classes.

In the case where n is even, the conjugacy classes are given by,

$$
\begin{aligned}
G = \{e\} \cup \{r, r^{n-1}\} \cup \{r^2, r^{n-2}\} \ldots \qquad (12.31)\\
\ldots \cup \{r^{(n-2)/2}, r^{(n+2)/2}\}\\
\cup \{r^{n/2}\}\\
\cup \{cr^{even}\} \cup \{cr^{odd}\}.
\end{aligned}
$$

In this case where n is even, there are a total of $1+\frac{(n-2)}{2}+1+2$ conjugacy classes.

12.5   Conjugacy Classes of the Quaternion Group

In this section we take a look at the conjugacy classes of the quaternion group, Q_8. Recall that the quaternion group is given by,

$$Q_8 = \{\pm 1, \pm i, \pm j, \pm k\}, \tag{12.32}$$

where the multiplicative structure is as described in Chapter (7). $Z(Q_8) = \pm 1$ hence each forms a singleton conjugacy class. Next we evaluate $x \notin Z(Q_8)$ in turn. We have,

$$jij^{-1} = ji(-j) = -jk = -i. \tag{12.33}$$

Similarly,

$$kik^{-1} = ki(-k) = -jk = -i. \tag{12.34}$$

It follows that

$$C_i = \{\pm i\} \tag{12.35}$$

The same holds for j and k, i.e.,

$$C_j = \{\pm j\} \quad \text{and} \quad C_k = \{\pm k\}. \tag{12.36}$$

Putting it together we get,

$$Q_8 = \{+1\} \cup \{-1\} \cup \{\pm i\} \cup \{\pm j\} \cup \{\pm k\}. \tag{12.37}$$

12.6   Conjugacy Classes of the Symmetric Group

The non-trivial elements of the symmetric group are products of cycles. The conjugacy classes are formed by the various cycle types. This follows by the conjugation rule, which states that conjugation of some $\tau \in S_n$ by some $\sigma \in S_n$ simply yields a relabeling of the elements of τ by σ, preserving the cycle structure. Therefore all elements of a certain cycle type belong to the same conjugacy class. The conjugation rule is written as,

$$\sigma(a_1 a_2 a_3 \ldots a_k)\sigma^{-1} = (\sigma(a_1)\sigma(a_2)\sigma(a_3)\ldots\sigma(a_k)). \quad (12.38)$$

One can easily check that this is consistent. For instance, let $\tau : a_i \to a_j$, then

$$\sigma\tau\sigma^{-1}(\sigma(a_i)) = \sigma\tau(a_i) = \sigma(a_j). \quad (12.39)$$

We leave the general proof as an exercise.

12.7   Exercises

1. Prove the orbit-stabilizer theorem.

2. Show why under the group action of conjugation, the centralizer is analogous to the stabilizer.

3. Prove that conjugacy classes are equivalence classes.

4. Consider the action of G on a set X, such that some x,y are conjugate to each other. Show that the stabilizer of x in G and the stabilizer of y in G are also conjugate to each other.

5. Derive the class equation.

6. Derive the version of the class equation in which conjugation is done by elements of a subgroup H of G, and acts on a set X.

7. List all the conjugacy classes of a group of order 43.

8. List all the conjugacy classes of a group of order 25.

9. Can a group of order 25 be non-abelian. Why or why not?

10. List all the conjugacy classes of a group of order 121.

11. Can a group of order 121 be non-abelian. Why or why not?

12. Consider D_{10}, the dihedral group of order 10. i.e. the symmetry group of the regular pentagon.

    a) How many conjugacy classes does it have?

    b) List all its conjugacy classes.

13. Consider D_{12}, the dihedral group of order 12, i.e. the symmetry group of the regular hexagon.

    a) How many conjugacy classes does it have?

b) List all its conjugacy classes.

14. What explains the difference between the number and types of conjugacy classes of D_{10} and D_{12}?

15. List all the conjugacy classes of S_3, the symmetric group on three letters.

16. Provide a general proof of the conjugation rule. i.e. show that conjugation preserves cycle structure.

17. Consider A_4, the alternating group on 4 letters.

    a) How many conjugacy classes does it have.

    b) Describe its conjugacy classes and state how many elements each contains.

18. Consider A_5, the alternating group on 5 letters.

    a) How many conjugacy classes does it have.

    b) Describe its conjugacy classes and state how many elements each contains.

19. Consider A_6, the alternating group on 6 letters.

    a) How many conjugacy classes does it have.

    b) Describe its conjugacy classes and state how many elements each contains.

20. Consider D_8, the dihedral group of order 8

    a) How many subgroups does it have?

    b) How many conjugacy classes of elements does it have?

    c) How many conjugacy classes of subgroups does it have?

21. Consider Q_8, the quaternion group.

    a) How many subgroups does it have?

b) How many conjugacy classes of elements does it have?

c) How many conjugacy classes of subgroups does it have?

22. What is the essential property of Q_8 that makes your answer for the above two questions different.

23. Show that all members of a conjugacy class have the same order. i.e. conjugation preserves order of elements.

24. Show that conjugation by any element of a group G is an automorphism, i.e. an isomorphism between G to itself.

25. Let $\mu, \sigma_1, \sigma_2 \in S_n$. Show that if σ_1 and σ_2 are disjoint, then so are $\mu\sigma_1\mu^{-1}$ and $\mu\sigma_2\mu^{-1}$.

Chapter 13

Class Equation Applications

$$|G| = |Z(G)| + \sum_{x \ rep} |C_x|$$

Tree Class

In class she sat, upon a mat, with its leaves above to shield the blast;
With sticks and stones, she added, subtracted, divided, high-lighted;
This Iroko tree of a class, will someday be a thing of the past;
But its lessons, roots, and strength of timber, will in her mind, forever last.

–S.G. Odaibo

The class equation though simple, is of great importance and has far reaching and diverse manifestations. It underpins several deep concepts in finite group theory and in Algebra in general. Here we present a number of key algebraic concepts and results as manifestations of the class equation. They include Cauchy's theorem; the abelian-ness of any finite group of order p^2; the non-triviality of the center of groups of order p^k where p is a prime; and the simplicity of the alternating group on five letters–A_5. Even the famous and essential Sylow theorems –presented in the next chapter– can be interpreted as a manifestation of the class equation.

Given a group G, the class equation is given by

$$|G| = |Z(G)| + \sum_{x \ rep} |C_x| \qquad (13.1)$$

where $Z(G)$ is the center of the group and C_x are the conjugacy classes. "$x \ rep$" denotes that the sum is over one representative element from each class, and where the elements in each C_x are not in $Z(G)$. The conjugacy classes are the orbits under the group action of conjugation.

13.1 $|Z(G)| \neq 1$ if $|G| = p^k$ where p prime.

Theorem 13.1. *If* $|G| = p^k$ *where* p *is a prime and* k *a natural number, then* $1 < |Z(G)| \leq p^k$. *i.e.* $|Z(G)| = p^j$ *for some* j *such that* $1 \leq j \leq k$.

Proof: By way of contradiction, assume $|Z(G)| = 1$. Then since p divides $|G|$ and p does not divide $|Z(G)|$, it follows from the class equation that p does not divide $\sum |C_x|$. Therefore there must exist at least one conjugacy class, C_{x_0}, whose order p does not divide.

$$p \nmid |C_{x_0}|. \qquad (13.2)$$

Therefore,

$$p \nmid \frac{|G|}{|C(x_0)|}. \tag{13.3}$$

This implies,

$$p \nmid \frac{p^k}{|C(x_0)|}. \tag{13.4}$$

This in turn implies,

$$|C(x_0)| = p^k = |G|. \tag{13.5}$$

And therefore since $C(x_0) \subset G$, it follows that $C(x_0) = G$. This is a contradiction, as $x_0 \notin Z(G)$. $\boxed{\text{QED}\checkmark}$

13.2   If $|G| = p^2$, G is abelian.

Theorem 13.2. *If $|G| = p^2$ where p is a prime, then G is abelian.*

Proof: By Theorem (13.1) above –stating that if the order of G is the power of a prime, then the center is non-trivial – it follows that $Z(G)$ here is non-trivial. Specifically, $1 < Z(G) \leq p^2$. By way of contradiction, assume G is non-abelian. Then

$$|Z(G)| \neq |G| = p^2. \tag{13.6}$$

Therefore is must be the case that

$$|Z(G)| = p. \tag{13.7}$$

Note that for any group G, the center, $Z(G)$, is contained in the centralizer, $C(x)$, of any group element x. i.e.

$$Z(G) \subset C(x). \tag{13.8}$$

Therefore,

$$|Z(G)| \leq |C(x)|. \tag{13.9}$$

Under our current assumption, it then follows that,

$$p \le |C(x)|. \tag{13.10}$$

Since $|Z(G)| \neq |G|$ there exists some $x_0 \notin Z(G)$. For such x_0,

$$C(x_0) \neq G, \tag{13.11}$$

Therefore

$$|C(x_0)| < |G|. \tag{13.12}$$

Also, for such x_0

$$|Z(G)| < |C(x_0)|. \tag{13.13}$$

Combining the above equations yields the following inequality,

$$p = |Z(G)| < |C(x_0)| < |G| = p^2, \tag{13.14}$$

i.e.,

$$p < |C(x_0)| < p^2. \tag{13.15}$$

However, since $C(x_0)$ is a subgroup of G, it follows by Lagrange's theorem that $|C(x_0)|$ must divide $|G|$, and hence the only allowable values for $|C(x_0)|$ are 1, p, and p^2. This is a contradiction. Hence any group G of order p^2 where p is a prime is abelian. $\boxed{\text{QED}\checkmark}$

13.3 Cauchy's Theorem

Theorem 13.3. *If p divides $|G|$ where p is a prime, then there exists some $x \in G$ such that $|x| = p$.*

Proof: We proceed by induction. Assuming Cauchy's theorem is true for $|G| \le n$, we need to show that it is true for $|G| = n + 1$. It is trivial to show that Cauchy theorem is true for $|G| = 2$ and $|G| = 3$ since both groups are of prime order and hence cyclic.

There are two main cases to consider:

Case I: $p\big||Z(G)|$. If $|Z(G)| = p$ then we are done with the proof, since every group of prime order is cyclic. If $|Z(G)| > p$, then pick any $x \in Z(G)$ such that $x \neq e$. Consider the subgroup H generated by x, i.e. $H = \langle x \rangle = \{x, x^2, x^3, \ldots, x^c\}$, where $c = |x|$. Note that if G is not abelian, then $Z(G) \neq G$, therefore there exists $y \in G$ such that $y \notin Z(G)$. And since $H \subset Z(G)$, it follows that $y \notin H$. Therefore $|H| < |G|$. Now if p divides $|H|$, then because $|H| < |G|$, by the induction premise we are again done with the proof. If however, $p \nmid |H|$, we will exploit the fact that $H \vartriangleleft G$. This is true because $H \subset Z(G)$, and therefore every element in H commutes with every element in G. Therefore,

$$gH = Hg \tag{13.16}$$

and hence,

$$gHg^{-1} = H. \tag{13.17}$$

Therefore

$$H \vartriangleleft G. \tag{13.18}$$

This allows one to form the quotient group G/H. Recall that the coset space is a group if the divisor group is normal. Of note,

$$|G/H| < |G|. \tag{13.19}$$

Furthermore,

$$p\big||G/H|, \tag{13.20}$$

because $|G/H| = |G|/|H|$ and $p \nmid |H|$. Therefore by the induction premise, it follows that there exists a subgroup K

of G/H such that $|K| = p$. Let $K = \langle yH \rangle$ for some $y \in G$, then since $|K| = p$, it follows that $y^p \in H$. If $y^p = e$, then $|y| = p$ and we are done with the proof. If on the other hand, $y^p \neq e$, then since $y^p \in \langle x \rangle$, it follows by Lagrange that $|\langle y^p \rangle|$ divides $|x|$. i.e. $|\langle y^p \rangle|$ divides c. This implies that $|\langle y^p \rangle| = c/m$ for some $m \in \mathbb{N}$. Therefore $|y^{c/m}| = p$, and we are done with the proof.

Case II: $p \nmid |Z(G)|$. Then since by the class equation,

$$p \big| \big(|G| = |Z(G)| + \sum |C_x| \big) \qquad (13.21)$$

it follows that

$$p \nmid \sum |C_x|. \qquad (13.22)$$

Therefore there exists some $x_0 \notin Z(G)$ such that

$$p \nmid |C_{x_0}| \qquad (13.23)$$

i.e.

$$p \nmid \frac{|G|}{|C(x_0)|}. \qquad (13.24)$$

This implies,

$$p \big| |C(x_0)|. \qquad (13.25)$$

We know

$$|C(x_0)| < |G| \qquad (13.26)$$

is true because $x_0 \notin Z(G)$. If $C(x_0) = G$, then $x_0 \in Z(G)$, which is not the case. Our induction premise therefore applies, and we know $C(x_0)$ contains and element x of order p. Such $x \in G$ also, because $C(x_0) \subset G$. This concludes the proof. $\boxed{\text{QED}\checkmark}$

In the above proof, the choice to split into the two cases — $p \big| |Z(G)|$ and $p \nmid |Z(G)|$ — was judicious and served a dual purpose. In the case $p\,|\,|Z(G)|$ we are guaranteed a non-trivial center. Therefore when we subsequently considered the quotient group G/H, it was guaranteed to be of smaller size than G, and hence the induction premise applied. Additionally, normality of the subgroup H was naturally inherited from $Z(G)$, thereby conferring group structure on the G/H coset space. In the second case, the mechanism of the proof is manifestly dependent on the non-divisibility by p as shown above.

13.4 A_5 is Simple

Theorem 13.4. *The Alternating group on 5 letters, A_5, is a simple group.*

Where *simple* means that the only normal subgroups are the trivial subgroup $\{e\}$ and the entire group.

Proof: Since conjugacy preserves cycle structure, the conjugacy classes of any alternating group are represented by the various cycle types it admits. These are the even permutations of n letters and all compositions of such permutations. In the case of A_5, representatives of the admissible cycle types are:

e, (123), (12)(34), and (12345).

A_5 is therefore a disjoint union of the associated conjugacy classes. i.e.

$$A_5 = \{e\} \bigcup C_{(123)} \bigcup C_{(12)(34)} \bigcup C_{(12345)}. \qquad (13.27)$$

Using the counting formula,

$$|C_x| = \frac{n!}{k(n-k)!}, \qquad (13.28)$$

where n is the number of letters (in this case, $n = 5$ since A_5) and k is cycle length. For example for (123), $k = 3$. The counting formula tells us how many elements are in each conjugacy class. Therefore since A_5 is a disjoint union of its conjugacy classes, the number of elements in A_5 is simply the sum of the number of elements in the conjugacy classes. The counting formula is that for "n pick k" where sequential order matters, but absolute order does not, hence the division by k. For instance, (123), (312), and (231) are all the same element in cycle notation, hence one must divide by 3. They however represent a different permutation from (132), (321), and (213), hence we do not divide by $3! = 6$. Of note, the counting formula requires modification where appropriate. For instance, the number of elements in $C_{(12)(34)}$ is given by

$$|C_{(12)(34)}| = \frac{(5 \cdot 4) \cdot (3 \cdot 2)}{2 \cdot 2 \cdot 2} = \frac{20 \cdot 6}{8} = 15. \qquad (13.29)$$

This is because there are 5 options for picking the first letter, 4 options for picking the second letter, 3 options for picking the third letter and 2 options for picking the fourth letter. One must then note that (12) and (21) represent the same permutation. Similarly, (12)(34) and (34)(21) represent the same permutation. Hence one must divide by 2 for the (12)-type element, divide again by 2 for (34)–the other transposition element, and divide a third time by 2 for the (12)(34) type element.

Proceeding with the formula and appropriate modifications were indicated, we get the following table:

Table 13.1: Conjugacy classes of A_5

Cycle type, x	$|C_x|$
$\{e\}$	1
(123)	20
(12)(34)	15
(12345)	$12 + 12$
	Total $= 60$

By definition of *normal*, $N \lhd G$ implies for any $g \in G$, $gNg^{-1} = N$. This implies for any $n \in N$, and for any $g \in G$, $gng^{-1} \in N$. Since the conjugacy class of n is defined by $C_n := \{gng^{-1}$ such that $g \in G\}$, it follows that for any $N \lhd G$, conjugacy classes of each $n \in N$ are contained in N in their entirety. And N is a disjoint union of its conjugacy classes.

As a consequence of the preceding, we have three constraints that must be satisfied by any $N \lhd G$:

1. $|N| = $ a sum of some combination of the orders of the conjugacy conjugacy classes of G.

2. N must include $\{e\}$, hence the combination being summed above must include 1.

3. Since N is a subgroup of G, by Lagrange, $|N|$ must divide $|G|$.

A careful inspection of Table (13.1) reveals that the only possible subgroups of G that can satisfy the above three constraints are $\{e\}$ and G itself. This concludes the proof that A_5 is a simple group. $\boxed{\text{QED}\checkmark}$

The above proof is direct and intuitive. It partly illuminates the forces that determine simplicity of a group. In particular, it sheds some light on the competition between the requirements for being a *group* –i.e. closure under inverse and closure under multiplication – and the requirement for being a *normal* subgroup – i.e. wholly containing the entire conjugacy class of each element in the subgroup. The requirements for a group are inviolate and hence the stronger of the two competing forces. A group is *simple* whenever the requirements for being a group disallow the satisfaction of the requirement for being a *normal* subgroup.

13.5 Exercises

1. Show that if G is an abelian group, every of its conjugacy classes are singletons.

2. Show that conjugacy preserves order of elements, i.e. any two elements in the same conjugacy class have the same order.

3. Show that conjugacy is an automorphism.

4. Given a group G and a subgroup H, show that the coset space is a group if and only if $H \triangleleft G$.

5. Tabulate the conjugacy classes of A_6 and calculate their respective orders.

6. Show that there are two conjugacy classes for the 5-cycle in A_5.

7. How many conjugacy classes are there for the 5-cycle in S_5 and why is this number different from that in A_5?

8. Prove that A_6 is simple.

9. Show that if $H \triangleleft G$ and $K \triangleleft G$, then $H \cap K \triangleleft H$ and $H \cap K \triangleleft K$.

10. Let G be a group of order 83521. Show that the center of G is non-trivial.

11. Explain why any group of order 77 contains a subgroup isomorphic to $\mathbb{Z}/11$.

12. Prove that any group of prime power order contains a non-trivial center.

13. Prove Cauchy's theorem.

14. Derive the conjugacy classes of the Quaternion group, Q_8.

15. Derive the conjugacy classes of the generalized Quaternion group, Q_{2n}.

16. Write all conjugacy classes of D_8.

17. Derive the conjugacy classes of the Dihedral group, D_{2n}.

18. Find all normal subgroups of D_5.

19. Find all normal subgroups of D_6.

20. Find all normal subgroups of D_n.

Chapter 14

The Sylow Theorems

The Silo

The silo, sticking out like a light house on a great plain,
Its grain is gain for months of pain,
A ray of sun, a gush of wind, this farm of mine, so mighty
fine,
This silo, prime. An hour glass that marks the time.

–S.G. Odaibo

The sylow theorems can be thought of as being along the same line as Cauchy's theorem, only they are much more powerful and sophisticated. Based on the prime factors of a group, the sylow theorems guarantee the existence of subgroups of prime factor power order. They also provide constraints on the numbers of such subgroups. And they tell us some things about the properties and relationship between these subgroups. The sylow theorems have far reaching consequences and play a central role in the classification of simple finite groups. There are three sylow theorems in all, each of which we state and prove below.

14.1 First Sylow Theorem

Theorem: Let G be a group such that $|G| = \prod_i p_i^{k_i}$, where p_i are distinct primes and $k_i \in \mathbb{N}$. Then for each i there exists a subgroup H_i of G such that $|H_i| = p_i^{k_i}$. H_i is called a sylow p_i subgroup of G.

Proof: Consider $|G| = p^k m = n$ where $gcd(p, m) = 1$, i.e. p is prime to m. We will proceed by induction. Assume that the theorem is true for $|G| < n$. We seek to show that it is true for $|G| = n$.

Since

$$|G| = |Z(G)| + \sum_{x \ rep} |C_x| \qquad (14.1)$$

where x is a representative element from the class C_x and $x \notin Z(G)$. And since p divides $|G|$, we can consider two cases:

Case i: $p \,|\, |Z(G)|$. Therefore by Cauchy's theorem there exists a subgroup H of $Z(G)$ such that $|H| = p$. Note $H \triangleleft G$ since $H \subset Z(G)$. It follows that G/H is a quotient group and

$$|G/H| = p^{k-1} m < n = |G|. \qquad (14.2)$$

Therefore by induction there exists a subgroup $K \subset G/H$ such that $|K| = p^{k-1}$. Each element $k \in K$ is of the form $k = xH$ where $x \in G$. Since K a subgroup of G/H it follows from the correspondence theorem —which we proved in Chapter (3)— that there exists a subgroup $J \subset G$ such that

$$J = \{xh \in G | h \in H \text{ and } xH \in K\}. \qquad (14.3)$$

The order of J is given by,

$$|J| = |H| \cdot |K| = p \cdot p^{k-1} = p^k. \qquad (14.4)$$

This concludes this part of the proof.

Case ii: $p \nmid |Z(G)|$. Then

$$p \nmid \left(\sum_x |C_x| \right) \qquad (14.5)$$

where x is rep and $\notin Z(G)$. Therefore there exists at least one $x_0 \notin Z(G)$ such that,

$$p \nmid |C_{x_0}|. \qquad (14.6)$$

It follows that,

$$p \nmid \frac{|G|}{|C(x_0)|} \qquad (14.7)$$

where $C(x_0)$ is the centralizer of x_0. This implies,

$$|C(x_0)| = p^k l < p^k m = |G|, \qquad (14.8)$$

for some natural number $l < m$. The above inequality is guaranteed because $x_0 \notin Z(G)$. It follows from our induction premise that there exists a subgroup

$$H \subset C(x_0) \qquad (14.9)$$

such that $|H| = p^k$. This concludes our proof of the first Sylow theorem. $\boxed{\text{QED}\checkmark}$

Another Proof of the First Sylow Theorem

Theorem: Again, let G be a group such that $|G| = \prod_i p_i^{k_i}$,

where p_i are distinct primes and $k_i \in \mathbb{N}$. Then for each i there exists a subgroup H_i of G such that $|H_i| = p_i^{k_i}$. H_i is called a sylow p_i subgroup of G.

Proof: Let $|G| = p^k m$ where p and m are relatively prime. And let

$$X = \{\text{All subsets } H \text{ of } G \text{ such that } |H| = p^k\}. \quad (14.10)$$

Then

$$|X| = \binom{p^k m}{p^k} = \frac{p^k m!}{p^k!(p^k m - p^k)!}. \quad (14.11)$$

Expanding the above expression yields,

$$|X| = \frac{p^k m \cdot (p^k m - 1) \cdot (p^k m - 2) \ldots \cdot (p^k m - p^k + 1)}{p^k \cdot (p^k - 1) \cdot (p^k - 2) \cdots \cdot 1}$$

$$(14.12)$$

The above is the combinatorial formula for "N pick j" in which the order does not matter, and where in this case $N = p^k m$ and $j = p^k$. So we have "$p^k m$ pick p^k" in which the order does not matter. The numerator arises because there are $p^k m$ options for picking the first element, then $p^k m - 1$ options for picking the second element, and $p^k m - p^k + 1$ options for picking the p^kth element. The denominator is because all $p^k!$ permutations of any chosen subset of p^k letters all represent the same entity (order does not matter), so this "over-counting must be corrected by dividing by $p^k!$.

In the above equation, each subsequent numerator term and the associated subsequent denominator term are each one less than their respective precursors in the sequence. For instance,

$$\frac{p^k m - j}{p^k - j} \tag{14.13}$$

is one less than

$$\frac{p^k m - j + 1}{p^k - j + 1}. \tag{14.14}$$

These numerator-denominator pairs therefore maintain the same factor of p as does the first pair $\frac{p^k m}{p^k}$. Since

$$gcd(p, m) = 1, \tag{14.15}$$

$$gcd\left(\frac{p^k m}{p^k}, p\right) = 1, \tag{14.16}$$

i.e. the first term is prime to p. It follows that all the other terms are also prime to p. This in turn means that the entire term $|X|$ is prime to p.

$$p \nmid |X|. \tag{14.17}$$

Consider the group action of left translation,

$$G \cdot X : X \to X, \tag{14.18}$$

defined by,

$$g \cdot H = gH. \tag{14.19}$$

Then the class equation for this action yields,

$$|X| = \sum_{rep\ H} |O_H|, \tag{14.20}$$

where O_H is the orbit of H under the G-action, and the sum is over representatives from each orbit. Now, since $p \nmid |X|$, it follows that,

$$p \nmid \left(\sum_{rep\ H} |O_H|\right). \tag{14.21}$$

Therefore there exists some H_0 such that,

$$p \nmid |O_{H_0}| \tag{14.22}$$

And therefore,

$$p \nmid \frac{|G|}{|Stab_G(H_0)|}. \tag{14.23}$$

This implies that $|Stab_G(H_0)|$ must contain the same power of p as $|G|$. Hence,

$$|Stab_G(H_0)| = p^k l < p^k m. \tag{14.24}$$

For some natural number $l < m$. We know that $H_0 \subsetneq G$, therefore $|Stab_G(H_0)| < |G|$. To show that $|H_0| = |Stab_G(H_0)|$, consider the right coset space $Stab_G(H_0) \backslash G$. Then by the definition of the stabilizer, for any $h \in H_0$,

$$Stab_G(H_0)h \subset H_0. \tag{14.25}$$

Furthermore,

$$|Stab_G(H_0)h| = |Stab_G(H_0)|, \tag{14.26}$$

and by Equation (14.25),

$$|Stab_G(H_0)h| \leq |H_0| = p^k. \tag{14.27}$$

Combining Equations (14.24), (14.26), and (14.27) yields,

$$p^k l = |Stab_G(H_0)| = |Stab_G(H_0)h| \leq |H_0| = p^k. \tag{14.28}$$

This implies,

$$p^k l \leq p^k, \tag{14.29}$$

which implies,

$$l = 1. \tag{14.30}$$

Consequently,

$$|Stab_G(H_0)| = p^k. \tag{14.31}$$

And this concludes the proof. $\boxed{\text{QED}\checkmark}$.

Summary: This proof constructs the set X of all subsets of size p^k. Next, the size of set X is shown to be prime to p. This is the key that through the class equation, guarantees the existence of some orbit, O_{H_0}, whose size is also prime to p. The orbit-stabilizer counting formula then tells us that both G and the stabilizer of H_0 have size with the same power of p factor. i.e. their sizes are $p^k m$ and $p^k l$ respectively. The last part of the proof required showing that l must equal 1, which showed that the stabilizer of H_0 has order p^k and is the Sylow p subgroup whose existence we sought to prove. This last part is accomplished by establishing a sequence of inequalities based on the definition of the stabilizer.

14.2 Second Sylow Theorem

Theorem: Consider a group G with order $p^k m$, where $gcd(p, m) = 1$. Given any subgroup $U \subset G$ with order p^r where $r \leq k$, U is contained in a conjugate of a sylow p subgroup. In particular, —considering the cases $r = k$— all sylow p subgroups are conjugate to one another.

Proof:
Let J be the coset space given by,

$$J = G/H \tag{14.32}$$

where H is some sylow p subgroup of G. Consider the group action of left translation acting on J.

$$G \cdot J : J \to J. \tag{14.33}$$

Given $j \in J$ such that $j = gH$, the above group action is defined by,

$$\tilde{g} \cdot j = \tilde{g}gH \in J, \tag{14.34}$$

where $\tilde{g}, g \in G$. Note that the group action is transitive. The stabilizer in G of the element $H \equiv e_J \in J$ is,

$$Stab_G(H \equiv e_J \in J) = H \subset G. \tag{14.35}$$

Therefore given the transitivity of this group action, for any $j \in J$,

$$Stab_G(j) = iHi^{-1} \tag{14.36}$$

where $j = iH$ for some $i \in G$. Since,

$$|iHi^{-1}| = |H|, \tag{14.37}$$

it follows that $Stab_G(j)$ is a sylow p subgroup for all $j \in J$.

$$|J| = |G|/|H| = \frac{p^k m}{p^k} = m. \tag{14.38}$$

Therefore,

$$p \nmid |J|. \tag{14.39}$$

Now consider the restriction of the above G-action to the subgroup $U \subset G$, where as noted above, $|U| = p^r$ with $r \leq k$. The class equation on J is,

$$|J| = \sum_{rep\ j} |\Theta_j|. \tag{14.40}$$

And because $p \nmid |J|$, it follows that,

$$p \nmid \sum_{rep\ j} |\Theta_j|. \tag{14.41}$$

This implies that there exists some $j_0 \in J$ such that,

$$p \nmid |\Theta_{j_0}|. \tag{14.42}$$

In light of this U-action on J, this implies,

$$p \nmid \frac{|U|}{|Stab_U(j_0)|}, \tag{14.43}$$

where,

$$Stab_U(j_0) = U \cap Stab_G(j_0). \tag{14.44}$$

Equation (14.43) thereby implies,

$$p \nmid \left(\frac{p^r}{|U \cap Stab_G(j_0)|} \right). \tag{14.45}$$

This in turn implies,

$$|U \cap Stab_G(j_0)| = p^r = |U|. \tag{14.46}$$

And therefore,

$$U \subset Stab_G(j_0) = iHi^{-1} \quad \text{for some } i \in G. \tag{14.47}$$

Therefore U is contained in some sylow p subgroup. And if $|U| = p^k$, then U is conjugate to H. This concludes the proof. $\boxed{\text{QED}\checkmark}$

Commentary: This proof proceeds in the same spirit as the proof of the First Sylow theorem in Subsection (14.1) above. Both proofs rely centrally on the class equation. In the first sylow theorem proof, the group action acted on the set X of all size p^k subsets of G. And $|X|$ was prime to p. Here, the group action is restricted to a p-subgroup, U, and acts on a coset space G/H where H is a sylow p subgroup. The order of the set acted on, i.e. $|G/H|$, is again prime to p. This yields us—via the class equation— that the orbit of some element is prime to p. This in turn —via the orbit stabilizer theorem— implies the order of the stabilizer in U must equal the order of U. This implies U must be contained in the stabilizer in U, which by transitivity of the group action, is conjugate to a sylow p subgroup.

14.3   Third Sylow Theorem

Theorem: Let G be a group such that $|G| = p^k m$ with m prime to p. Then regarding the number n_p of Sylow p subgroups, it holds that:

a) $n_p \mid m$, and

b) $n_p \equiv 1 (mod\ p)$.

Proof of Part (a): Since all sylow p subgroups are conjugate, it follows that under the group action of conjugation they all belong to the same orbit. In other words there is only one conjugacy class for the set S of sylow p subgroups. Given any sylow p subgroup H_p, the orbit-stabilizer counting formula reads,

$$|S| = \frac{|G|}{|Stab_G(H_p)|} \tag{14.48}$$

Now considering that under the group action of conjugation, the stabilizer of a subgroup is the normalizer, i.e.

$$|Stab_G(H_p)| = N(H_p), \tag{14.49}$$

it follows that,

$$|S| = \frac{|G|}{|N(H_p)|}, \tag{14.50}$$

where $N(H_p)$ is the normalizer of H_p. i.e.

$$N(H_p) = \{g \in G \mid gH_pg^{-1} = H_p\}. \tag{14.51}$$

Clearly

$$H_p \subset N(H_p). \tag{14.52}$$

And therefore by Lagrange,

$$|H_p| \mid |N(H_p)| \tag{14.53}$$

This implies that,

$$|N(H_p)| = p^k l. \tag{14.54}$$

Consequently,

$$n_p = |S| = \frac{p^k m}{p^k l} = \frac{m}{l}. \tag{14.55}$$

And since

$$\frac{m}{l} \mid m, \tag{14.56}$$

it follows that,

$$n_p \mid m. \tag{14.57}$$

This concludes the proof of part (a) of the theorem. $\boxed{\text{QED}\checkmark}$

Summary: The the key to the proof here is that all sylow p subgroups are conjugate to each other, and therefore there is only a single orbit under the group action of conjugation. The resulting orbit-stabilizer counting formula directly yields the proof that $n_p \mid m$.

Proof of Part (b): Again let S be the set of all sylow p subgroups,

$$S = \{H \mid H \text{ is a sylow } p \text{ subgroup of } G\}. \qquad (14.58)$$

Consider the restriction of the conjugation group action to H.

Claim: H is the only Sylow p subgroup fixed under this action. i.e.

$$Stab_H(H) = H \qquad (14.59)$$

and

$$Stab_H(H') \subsetneq H \qquad (14.60)$$

for every $H' \neq H$. Here, the class equation reads,

$$|S| = \frac{|H|}{|Stab_H(H)|} + \sum_{rep\ H' \neq H} \frac{|H|}{|Stab_H(H)|}. \qquad (14.61)$$

The claim therefore implies,

$$|S| = 1 + \sum_{rep\ H' \neq H} \frac{p^k}{p^r} \qquad (14.62)$$

where $0 \leq r < k$. Therefore,

$$n_p = |S| = 1 + \sum_i p^{k - r_i} \qquad (14.63)$$

where $0 \leq r_i < k$.

Therefore n_p is the sum of 1 and a term which is a power of p. In other words,

$$n_p \equiv 1(mod\ p). \qquad (14.64)$$

This shows that if the above claim is true, then the theorem is true. All that remains now is to prove the claim, after which we are done with the proof of the theorem.

Proof of Claim: By way of contradiction, assume the above H-action fixes some Sylow p subgroup $H' \neq H$. Then,

$$hH'h^{-1} = H' \qquad (14.65)$$

for all $h \in H$. This implies

$$H \subset N(H'). \qquad (14.66)$$

However, it is also clearly true that,

$$h'H'h'^{-1} = H' \qquad (14.67)$$

for all $h' \in H'$.
Hence it follows also that,

$$H' \subset N(H'). \qquad (14.68)$$

Therefore H and H' and both subgroups of $N(H')$, and therefore are both Sylow p subgroups of $N(H')$. Hence according to the Second Sylow Theorem, H and H' are conjugate to one another in $N(H')$. i.e.

$$H = n_0 H' n_0^{-1} \qquad (14.69)$$

for some $n_0 \in N(H')$. However, by definition of the normalizer,

$$N(H') = \{g \in G \mid gH'g^{-1} = H'\}, \qquad (14.70)$$

$$gH'g^{-1} = H' \text{ for all } g \in N(H'), \qquad (14.71)$$

and

$$H' \triangleleft N(H'). \qquad (14.72)$$

The above three equations all contradict Equation (14.69). This concludes the proof. $\boxed{\text{QED}\checkmark}$

Summary: The statement $n_p \equiv 1 (mod\ p)$ is equivalent to the statement that the sylow p subgroup H is the solitary fixed point of the conjugation group action restricted to H and acting on the set of all sylow p subgroups. This is the key to the proof. The proof first establishes this equivalence and then proceeds to prove it. The equivalence is established by simply substituting the appropriate orbit-stabilizer counting formula into the class equation. The proof is then completed by way of contradiction as follows: If the above conjugation H-action is assumed to also fix some other Sylow p subgroup H', then both H and H' are in $N(H')$. They are therefore both sylow p subgroups of $N(H')$, and are therefore conjugate to each other. This is a contradiction to $H' \lhd N(H')$. And so goes this truly elegant proof.

14.4 Exercises

1. Let H and H' be sylow p subgroups of G. Show that $N(H) \cap N(H')$ cannot contain either sylow p subgroup.

2. Apply sylow theory to A_4, i.e. use sylow theory to determine the numbers and types of its subgroups.

3. List all Sylow subgroups of A_5. Are any of the sylow subgroups normal?

4. Apply sylow theory to S_5.

5. Apply sylow theory to a group G of order 30.

6. Apply sylow theory to a group of order 77.

7. Derive the formula for "N pick M" where order does not matter.

8. Prove the first Sylow theorem.

9. Prove the second Sylow theorem.

10. Prove the third Sylow theorem.

11. Prove that there are no simple groups of order 48.

12. Prove that there are no simple groups of order 200.

13. Prove that there are no simple groups of order 192.

14. Prove that there are no simple groups of order 500.

15. Prove that there are no simple groups of order 280.

16. Prove that there are no simple groups of order 160.

17. Prove that there are no simple groups of order 1452.

18. Show that all groups of order p^2q have a normal sylow p or a normal sylow q subgroup. Assume p and q are distinct primes with $p < q$.

19. Let G be a group for which every of its Sylow subgroups are normal. Show that G is isomorphic to the direct product of its Sylow subgroups.

20. Consider primes $p < q$ for which $p \not\equiv 1(mod\ q)$. Show that every group of order pq is cyclic.

21. Let U be a subgroup of G such that $|U| = p^r$, where p is prime and $r \in \mathbb{N}$. Show that U is contained in some sylow p subgroup.

Chapter 15

Applications of the Sylow Theorems

15.1 $|G| = pq$

Here we analyze the case $|G| = pq$, where p and q are primes and $q > p$.

Theorem 15.1. *The sylow q subgroup is characteristically normal in G, i.e. $H_q \lhd G$.*

Proof: From the sylow theorems, the number of sylow q subgroups, n_q adheres to the following constraints,

$$n_q \equiv 1 (mod\ q) \tag{15.1}$$

and

$$n_q \mid p. \tag{15.2}$$

Since $q > p$, 1 is the only member of $1(mod\ q)$ that can divide p. Therefore $n_q = 1$, and $H_q \lhd G$. $\boxed{\text{QED}\checkmark}$

Theorem 15.2. *If $p \nmid q - 1$ then $H_p \lhd G$, i.e. $n_p = 1$.*

Proof: By the sylow theorems,

$$n_p \mid q, \tag{15.3}$$

therefore the two possible values of n_p are $n_p = 1$ and $n_p = q$. Also by Sylow,

$$n_p \equiv 1 (mod\ p) \tag{15.4}$$

i.e.,

$$n_p = kp + 1 \tag{15.5}$$

for some natural number k. Rewriting we get,

$$n_p - 1 = kp. \tag{15.6}$$

Next we substitute $n_p = q$ into the above equation to check its implication:

$$q - 1 = kp \tag{15.7}$$

And this implies,

$$p \mid q - 1. \tag{15.8}$$

Working backwards, we see that the converse is also true that if,

$$p \nmid q - 1, \tag{15.9}$$

then for any $k \in \mathbb{N}$,

$$q - 1 \neq kp. \tag{15.10}$$

Hence,

$$q \neq kp + 1, \tag{15.11}$$

i.e.

$$q \not\equiv 1 (mod\ p) \tag{15.12}$$

therefore,

$$n_p \neq q \tag{15.13}$$

therefore,

$$n_p = 1 \tag{15.14}$$

i.e.

$$H_p \triangleleft G \tag{15.15}$$

$\boxed{\text{QED}\checkmark}$

On Generators of Groups of Order pq

Let G be a group of order pq, such that p and q are primes and $q > p$. By Sylow theory there exists some subgroup H_q of order q, and subgroup H_p of order p. Every element in G is of the form xy where $x \in H_p$ and $y \in H_q$. To see this note that you can form the quotient group G/H_q which contains p elements of the form $x_p H_q$. And since $H_q \lhd G$, G/H_q is a group. Then by the lattice theorem,

$$G = \{xy | x \in H_p, y \in H_q\} \qquad (15.16)$$

Then given any $g = xy$ and $g' = x'y' \in G$, we require gg' in terms of elements of H_p and H_q.

$$gg' = (xy)(x'y'). \qquad (15.17)$$

Note that since p and q prime, H_p and H_q are cyclic and given by,

$$H_p = \langle x \rangle = \{x, x^2, x^3, ..., x^p\} \qquad (15.18)$$

and

$$H_q = \langle y \rangle = \{y, y^2, y^3, ..., y^q\} \qquad (15.19)$$

Specifically, what we require is a solution to the following for j,

$$yx' = x'y^j \qquad (15.20)$$

where $0 < j < q$. Having such a solution will enable us write Equation (15.17) as,

$$gg' = (xx')(y^j y') \qquad (15.21)$$

where $xx' \in H_p$ and $y^j y' \in H_q$. The above commutation equation can be recast as,

$$x'^{-1} yx' = y^j. \qquad (15.22)$$

In this form, we readily see that a solution j always exists since $H_q \lhd G$. Conjugation is a homomorphism, therefore conjugating both sides of the above equation yields,

$$x'^{-1}x'^{-1}yx'x' = x'^{-1}y^j x' = \underbrace{(x'^{-1}yx')(x'^{-1}yx')\ldots}_{j\ times} \quad (15.23)$$

Consolidating and applying Equation (15.22) yields,

$$x'^{-2}yx'^2 = x'^{-1}y^j x' = \underbrace{y^j y^j \ldots}_{j\ times} \quad (15.24)$$

$$x'^{-2}yx'^2 = y^{j^2} \quad (15.25)$$

We readily see from the above that if we conjugate p times we get,

$$x'^{-p}yx'^p = y^{j^p} \quad (15.26)$$

Note that

$$x'^{-p} = x'^p = e, \quad (15.27)$$

Therefore we get,

$$x'^{-p}yx'^p = y = y^{j^p} \quad (15.28)$$

More generally,

$$y^{1(mod\ q)} = y^{j^p} \quad (15.29)$$

Therefore our *general solution* is that j must satisfy the condition,

$$j^p \equiv 1(mod\ q) \quad (15.30)$$

This condition determines which groups of order pq exist for any given p and q prime. For example, consider $|G| = 2 \cdot 3 = 6$. Here, $p = 2$ and $q = 3$.

$$H_q \simeq \mathbb{Z}/3 = (\{0, 1, 2\}, +) \tag{15.31}$$

We see that the values $j = 1$ and $j = 2$ both satisfy the pq condition $j^p \equiv 1 (mod\ q)$. See that,

$$1^2 = 1 \equiv 1 (mod\ 3) \tag{15.32}$$

and

$$2^2 = 4 \equiv 1 (mod\ 3) \tag{15.33}$$

Note that regardless of the values of p and q, 1^p is always congruent to $1 (mod\ q)$. Therefore there is always a commutative group of order pq. In particular, the cyclic group $\mathbb{Z}/pq$ always exists for any p and q. Also, when $p = 2$, note that $(q - 1)^p$ is always congruent to $1 (mod\ q)$ since

$$(q - 1)^2 = q^2 - 2q + 1 = kq + 1 \equiv 1 (mod\ q) \tag{15.34}$$

where $k = q - 2 \in \mathbb{N}$. Therefore $j = q - 1$ is always a solution when $p = 2$. Note that in $\mathbb{Z}/q$, 1 is the inverse of $q - 1$. Therefore $y^{-1} \equiv y^{q-1}$. Hence the commutation relationship, Equation (15.22) arising out of the second case above is,

$$x'^{-1}yx' = y^{-1}. \tag{15.35}$$

We immediately recognize this as the defining relation of the Dihedral group of order pq.

Theorem 15.3. *Every group of order* $2q$*, where* q *is a prime greater than 2, is isomorphic to either the cyclic group* $\mathbb{Z}/2q$ *or the Dihedral group of order* $2q$*.*

Proof: We have already shown above that the commutation $x'y = x'y^j$ of such a group must satisfy the following condition:

$$j^2 \equiv 1 (mod \ q) \tag{15.36}$$

Furthermore we showed that this condition is always satisfied by $j = 1$ and by $j = q - 1$. Also, we showed that the $j = 1$ case yields the cyclic group of order pq, while the $j = q - 1$ case yields the dihedral group of order pq. To complete the proof, all that remains is to show that the commutation condition cannot be satisfied by any other $j \in \mathbb{Z}/q$, and therefore no other group isomorphisms are admissible. Consider some $j \in \mathbb{Z}/q$ such that $j \neq 1$ and $j \neq q - 1$. Then $j = q - l$ for some $l \neq 1$ and $l \neq q - 1$.

$$j^2 = (q - l)^2 = q^2 - 2ql + l^2 = q(q - 2l) + l^2. \tag{15.37}$$

The above expression is congruent to $1 (mod \ q)$ if and only if l^2 is congruent to $1 (mod \ q)$. By way of contradiction, assume

$$l^2 \equiv 1 (mod \ q). \tag{15.38}$$

Then,

$$l^2 - 1 = kq \tag{15.39}$$

for some $k \in \mathbb{N}$. Then,

$$q \mid (l - 1)(l + 1) \tag{15.40}$$

This is a contradiction because it requires q to appear in the prime number factorization of either $(l-1)$ or $(l+1)$. This cannot happen since $q > (l - 1)$ and $q > (l + 1)$. $\boxed{\text{QED}\checkmark}$

Example 15.4. For our next example, we examine the case $|G| = 3 \cdot 5 = 15$. Here $p = 3$ and $q = 5$.

Here, the commutation $x'y = x'y^j$ of such a group must satisfy the following condition:

$$j^3 \equiv 1 (mod\ 5) \tag{15.41}$$

Here we tabulate the non-trivial elements of $j \in \mathbb{Z}/q$ alongside the corresponding values of j^3. We get,

Table 15.1: Case $|G| = pq = 15$

j	j^3	mod 5
1	1	1
2	8	3
3	27	2
4	64	4

In the above table, the only admissible value of j is $j = 1$. None of the other values satisfy the condition of Equation (15.41). Therefore the only possibility is $x'y = x'y$, i.e. both generators commute. This is the cyclic group of order pq, and is the only isomorphism class for groups of order 15. In this example, $G \simeq \mathbb{Z}/15$ only because $5 \not\equiv 1 (mod\ 3)$. We next state and prove this claim more generally as a theorem.

Theorem 15.5. If $|G| = pq$ where p and q are primes with $p < q$ and $q \not\equiv 1 (mod\ p)$, then $G \simeq \mathbb{Z}/pq$

Proof: In Theorem (15.1), we proved that if $p \nmid q - 1$ then the sylow p subgroup is normal in G. Here we proceed by showing that $p \nmid q-1$ is equivalent to $q \not\equiv 1(mod\ p)$, hence $H_p \lhd G$. We then complete the proof by using the normality of H_p and H_q and the same line of reasoning that yielded us the commutation condition on groups of order pq above.

If $p \nmid q - 1$ then $q - 1 \neq kp$ for any nontrivial $k \in \mathbb{Z}$. Therefore $q \neq kp + 1$ for any nontrivial $k \in \mathbb{Z}$. Hence $q \not\equiv 1(mod\ p)$. This implies $H_p \lhd G$. Since both $H_p = \langle x \rangle$ and $H_q = \langle y \rangle$ are normal in G and are cyclic, it follows that,

$$xyx^{-1} = y^j \tag{15.42}$$

for some $j \in \mathbb{Z}/q \setminus \{0\}$. Similarly,

$$y^{-1}xy = x^k \tag{15.43}$$

for some $k \in \mathbb{Z}/p \setminus \{0\}$.

It follows from the above two equations that,

$$xy = y^j x \tag{15.44}$$

and

$$xy = yx^k. \tag{15.45}$$

This implies that,

$$y^j x = yx^k. \tag{15.46}$$

Therefore,

$$y^{j-1} = x^{k-1}. \tag{15.47}$$

However, since

$$\langle y \rangle \cap \langle x \rangle = \{e\}, \tag{15.48}$$

It must be the case that

$$y^{j-1} = x^{k-1} = \{e\}. \tag{15.49}$$

And since $j < q$ and $k < p < q$, the only possibility is

$$j = k = 1 \tag{15.50}$$

This implies that

$$xy = yx \tag{15.51}$$

i.e. the generators commute, hence

$$G = \langle xy \rangle \tag{15.52}$$

This completes our proof that if $|G| = pq$ where p and q are primes with $p < q$ and $q \not\equiv 1 (mod\ p)$, then G is cyclic.

QED✓

Theorem 15.6. *If a group G is such that $|G| = p^r q^l$ where $r, l \in \mathbb{N}$ and $p \neq q$. And if $n_p = n_q = 1$, then elements of the sylow p and sylow q subgroups commute with each other.*

Proof: Proceeding in the same spirit as the above proof, we note that by Lagrange $H_p \cap H_q = \{e\}$. Also, since $H_p \triangleleft G$ and $H_q \triangleleft G$, it follows that for any $x \in H_p$ and $y \in H_q$

$$y^{-1}xy \in H_p \tag{15.53}$$

Therefore by closure under group multiplication,

$$(y^{-1}xy)x^{-1} \in H_p \tag{15.54}$$

However,

$$y^{-1}(xyx^{-1}) \in H_q \tag{15.55}$$

Therefore,

$$(y^{-1}xy)x^{-1} = y^{-1}(xyx^{-1}) \in H_p \cap H_q = \{e\} \tag{15.56}$$

Therefore,

$$xy = yx \tag{15.57}$$

QED✓

Theorem 15.7. *G equals the direct product of its sylow p subgroups if and only if each of its sylow p subgroups are normal in G.*

i.e.

$$G = \mathcal{P}_1 \times \mathcal{P}_2 \times \cdots \times \mathcal{P}_n \iff \mathcal{P}_i \triangleleft G \quad \forall\, i \qquad (15.58)$$

where $\mathcal{P}_i$ are the sylow p subgroups.

Proof: In part one of the proof, we show that if each sylow p subgroup is normal then G equals the direct product of its sylow p subgroups. In part two of the proof we show that if each G equals the direct product of its sylow p subgroups, then each of the sylow p subgroups must be normal.

Let each sylow p subgroup be normal in G. We will describe an isomorphism between G and $\mathcal{P}_1 \times \mathcal{P}_2 \times \cdots \times \mathcal{P}_n$. Let ϕ be

$$\phi : G \to \mathcal{P}_1 \times \mathcal{P}_2 \times \cdots \times \mathcal{P}_n \qquad (15.59)$$

$$\phi(g) = (e, e, \cdots \underbrace{g}_{(j)}, \cdots, \underbrace{e}_{(n)}) \qquad (15.60)$$

for $g \in G$ and $g \in \mathcal{P}_j$ for some $j \leq n$. Note that this accounts only for

$$1 + \sum_j (|\mathcal{P}_j| - 1) \qquad (15.61)$$

elements. The above is the identity plus the non-trivial elements of each sylow p subgroup. This set is the 'basis' of the group G. We will see the remaining elements are necessarily products of elements in this 'basis set'. Now given any element $s \in G$ such that s is a product of elements of different sylow p subgroups, e.g say $s = xy$ where $x \in \mathcal{P}_i$ and $y \in \mathcal{P}_j$, we define ϕ as acting as follows,

$$\phi(s) = (e, e, \cdots \underbrace{x}_{(i)}, \cdots, \underbrace{y}_{(j)}, \cdots, \underbrace{e}_{(n)}) \qquad (15.62)$$

From the above definition it follows that,

$$\phi(e) = (e, e, \cdots, \underbrace{e}_{(n)}) \tag{15.63}$$

Lemma 15.8. ϕ *is an isomorphism between G and $\mathcal{P}_1 \times \mathcal{P}_2 \times \cdots \times \mathcal{P}_n$*

Proof: We show onto, $1-1$, and homomorphism. For onto, note that given any arbitrary element $t \in \mathcal{P}_1 \times \mathcal{P}_2 \times \cdots \times \mathcal{P}_n$, we have,

$$t = (t_1, t_2, \cdots t_n) \tag{15.64}$$

where $t_j \in \mathcal{P}_j$ for each j. Therefore from the definition of ϕ,

$$\phi\left(\prod_{j=1}^{n} t_j\right) = (t_1, t_2, \cdots t_n) = t \tag{15.65}$$

and $\prod_{j=1}^{n} t_j \in G$ by closure under multiplication. Furthermore from definition of sylow p subgroup,

$$|G| = \prod_{j=1}^{n} |\mathcal{P}_j| \tag{15.66}$$

While by definition of the Cartesian space,

$$|\mathcal{P}_1 \times \mathcal{P}_2 \times \cdots \times \mathcal{P}_n| = \prod_{j=1}^{n} |\mathcal{P}_j| \tag{15.67}$$

Hence,

$$|G| = |\mathcal{P}_1 \times \mathcal{P}_2 \times \cdots \times \mathcal{P}_n|. \tag{15.68}$$

The cardinality match further proves to us that every element in G is decomposable into a product of elements of the different sylow subgroups.

For $1-1$, assume by way of contradiction that there exists some $u, v \in G$ such that $u \neq v$, but $\phi(u) = \phi(v)$. $u =$

$\prod_{j=1}^{n} u_j$ and $v = \prod_{j=1}^{n} v_j$, while $\phi(u) = (u_1, u_2, \cdots u_n)$ and $\phi(v) = (v_1, v_2, \cdots v_n)$, where $u_j, v_j \in P_j$. Since $u \neq v$, there must exists some j such that $u_j \neq v_j$. This is a contradiction since equality in the Cartesian space implies coordinate-wise equality.

For homomorphism, consider $g, h \in G$. Then $g = \prod_{j=1}^{n} g_j$, $h = \prod_{j=1}^{n} h_j$, $\phi(g) = (g_1, g_2, \cdots g_n)$, and $\phi(h) = (h_1, h_2, \cdots h_n)$, where $g_j, h_j \in P_j$. It follows from direct product definition that,

$$\phi(g) \cdot \phi(h) = (g_1 h_1, g_2 h_2, \cdots g_n h_n) \qquad (15.69)$$

On the other hand,

$$gh = \prod_{j=1}^{n} g_j \prod_{j=1}^{n} h_j \qquad (15.70)$$

By Theorem (15.6), since each of the different sylow subgroups of G are normal in G, their elements commute with each other. Hence it follows that,

$$gh = \prod_{j=1}^{n} g_j \prod_{j=1}^{n} h_j = \prod_{j=1}^{n} g_j h_j \qquad (15.71)$$

Therefore,

$$\phi(gh) = \phi\left(\prod_{j=1}^{n} g_j h_j\right) = (g_1 h_1, g_2 h_2, \cdots g_n h_n) \qquad (15.72)$$

Therefore,

$$\phi(gh) = \phi(g) \cdot \phi(h) \qquad (15.73)$$

This completes the first part of the proof. $\boxed{\text{QED}\checkmark}$

In the second part of the proof we show that if $G = P_1 \times P_2 \times \cdots \times P_n$ then each sylow p subgroup is normal. Let $G = P_1 \times P_2 \times \cdots \times P_n$. Then each pick an arbitrary element $x \in \mathcal{P}_j$. It follows that, $x = (e, e, \ldots, \underbrace{x}_{(j)}, \ldots, e)$.

Consider an arbitrary element $y \in G$. It follows that, $y = (y_1, y_2, \ldots, y_j, \ldots, y_n)$ and $y^{-1} = (y_1^{-1}, y_2^{-1}, \ldots, y_j^{-1}, \ldots, y_n^{-1})$. Conjugating x by y we get,

$$yxy^{-1} = (y_1 e y_1^{-1}, y_2 e y_2^{-1}, \ldots, y_j x y_j^{-1}, \ldots, y_n^{-1}) \quad (15.74)$$
$$= (e, e, \ldots, y_j x y_j^{-1}, \ldots, e)$$

It follows that if $x \in \mathcal{P}_j$, then $yxy^{-1} \in \mathcal{P}_j$ for any $y \in G$. Therefore $\mathcal{P}_j \triangleleft G$ for all j. This completes the proof. $\boxed{\text{QED}\checkmark}$

15.2 $|G| = p^2q$

Theorem 15.9. *If G is a group with $|G| = p^2q$, where p, q are primes and $p < q$, then G is solvable.*

Proof: We first review the definition of *solvable*, then we show that if $|G| = p^2q$ either the sylow p or the sylow q subgroup must be normal in G.

Definition 15.10. A group G is called *solvable* if it has some sequence of subgroups H_i such that $\{e\} \lhd H_1 \lhd H_2 \lhd \cdots \lhd H_{n-1} \lhd G$, and the quotient group H_{i+1}/H_i is abelian for each i in the sequence.

Lemma 15.11. *If G is a group such that $|G| = p^2q$ where p, q prime and $p < q$, then either the sylow p or the sylow q subgroup must be normal in G.*

Proof of Lemma: By way of contradiction assume neither the sylow p nor the sylow q subgroups are normal in G. Then $n_p > 1$ and $n_q > 1$. By the sylow theorem, $n_q \equiv 1 (mod\ q)$, i.e.,

$$n_q = kq + 1 \tag{15.75}$$

for some non-negative integer k. Furthermore,

$$n_q \mid p^2 \tag{15.76}$$

Since $p < q$ and $n_q > 1$, it follows that $n_q \nmid p$ and $n_q \nmid 1$. Therefore,

$$n_q = p^2 \tag{15.77}$$

Since q is a prime, the intersection of distinct sylow q subgroups is trivial, therefore there are $n_q(q - 1)$ elements of order q. i.e. there are $p^2(q - 1) = p^2q - p^2 = |G| - p^2$ elements of order q. There therefore is only room for one sylow p subgroup. Hence the sylow p subgroup is normal in

G, and this is a contradiction. This implies that either the sylow p or sylow q subgroups are normal in G. $\boxed{\text{QED}\checkmark}$

If $H_p \lhd G$, then $\{e\} \lhd H_p \lhd G$, and since $|H_p| = p^2$, by Theorem (13.2), H_p is abelian. Also, $G/H_p \simeq \mathbb{Z}/q$ therefore abelian. It follows that G is solvable. On the other hand, if we have $H_q \lhd G$, and $\{e\} \lhd H_q \lhd G$. Then since $H_q \simeq \mathbb{Z}/q$, H_q abelian. Also, $|G/H_q| = p^2$ hence abelian too. Therefore G is solvable. $\boxed{\text{QED}\checkmark}$

Theorem 15.12. *Let G be a group with $|G| = p^2q$ where p, q are primes and $p < q$. If $q \not\cong 1(mod\ p)$ and $p^2 \not\cong 1(mod\ q)$, then G is abelian.*

Proof: By Sylow, $n_p \cong 1(mod\ p)$ and divides q. Therefore since $q \not\cong 1(mod\ p)$, n_p cannot equal q. Hence $n_p = 1$. Similarly, $n_q \cong 1(mod\ q)$ and divides p^2. However because $p < q$, n_q does not divide p. Hence n_q either equals 1 or p^2. But since $p^2 \not\cong 1(mod\ q)$, n_q cannot equal p^2. Therefore $n_q = 1$. Hence by Theorem (15.6), the elements of the Sylow q and the Sylow p subgroups commute with each other. And since they have order prime and prime squared respectively, they are each abelian. Therefore any element in G commutes with any other element in G, hence G is abelian. $\boxed{\text{QED}\checkmark}$

15.3 If G simple of order 60, then **G** ≃ **A₅**

Theorem 15.13. *If G is a simple group of order* 60, *then* G *is isomorphic to* A_5

    Proof: The proof will proceed by accounting for the number of elements of a prime divisor order of G, i.e. 5, then 3, then 2. Along the way, establishing constraints for what types and numbers of elements can occupy the remaining space, and utilizing the fact that the total number of elements adds up to and cannot exceed 60. This will lead to a demonstration that G contains a subgroup of order 12. Quotienting G by this group yields a coset space of order 5. By Cayley's theorem, we will then show that G's left regular action on this coset space yields a subgroup of S_5. We then conclude the proof by showing that any subgroup of S_5 with index 2 is isomorphic to A_5.

$$|G| = 60 = 2^2 \cdot 3 \cdot 5 \tag{15.78}$$

Lemma 15.14. *If G is simple, then the number of sylow p subgroups, $n_p(G)$, is greater than one for all p.*

    Proof: This lemma follows from the Sylow theorems. According to the second Sylow theorem, all sylow p subgroups are conjugate to each other. If by way of contradiction we assume $n_p(G) = 1$ for some p; then for the associated sylow p subgroup H_p,

$$gH_pg^{-1} = H_p \tag{15.79}$$

for all $g \in G$. This implies,

$$H_p \lhd G, \tag{15.80}$$

which is a contradiction to the simplicity of G. $\boxed{\text{QED}\checkmark}$.

Lemma 15.15. *For any fixed prime factor p of G, the normalizers of the sylow p subgroups are conjugate and therefore isomorphic to each other.*

Proof: According to the second sylow theorem all sylow p subgroups are conjugate. Therefore given any H_p and H'_p such that

$$H_p \neq H'_p, \tag{15.81}$$

there exists some $g \in G$ such that,

$$gH_pg^{-1} = H'_p \tag{15.82}$$

Therefore,

$$H_p = g^{-1}H'_pg. \tag{15.83}$$

Hence given any arbitrary $h' \in H'_p$,

$$g^{-1}h'g \in H_p \tag{15.84}$$

Similarly, given arbitrary $h \in H_p$,

$$ghg^{-1} \in H'_p \tag{15.85}$$

Equation (15.83) implies that for any given p, conjugation is a transitive action on the set of sylow p subgroups. This therefore implies that the stabilizers under this group action are also conjugate to each other.

$$g(Stab_G(H_p))g^{-1} = Stab_G(H'_p) \tag{15.86}$$

Since the stabilizers under conjugation are the normalizers, the normalizers are conjugate to each other.

$$gN(H_p)g^{-1} = N(H'_p) \tag{15.87}$$

To verify this, consider an arbitrary $n \in N(H_p)$. Then,

$$n(H_p)n^{-1} = H_p \qquad (15.88)$$

and the conjugation by g transforms n into an element of $gN(H_p)g^{-1}$ as follows,

$$n \to gng^{-1} \qquad (15.89)$$

Conjugating an arbitrary element $h' \in H_p'$ by gng^{-1} yields,

$$(gng^{-1})h'(gng^{-1})^{-1} = (gng^{-1})h'(gn^{-1}g^{-1}). \qquad (15.90)$$

Regrouping by associativity, we get,

$$(gn)(g^{-1}h'g)(n^{-1}g^{-1}) = (gn)h(n^{-1}g^{-1}), \qquad (15.91)$$

where $h \in H_p$ and we have used Equation (15.84) to make the substitution above. Further regrouping by associativity we get,

$$(gn)h(n^{-1}g^{-1}) = g(nhn^{-1})g^{-1} = g\tilde{h}g^{-1} \in H_p', \qquad (15.92)$$

where $\tilde{h} \in H_p$ and we have used Equation (15.88), that $n \in N(H_p)$, and Equation (15.85), that $ghg^{-1} \in H_p'$ for all $h \in H_p$. Equations (15.90), (15.91), and (15.92) together imply,

$$(gng^{-1})h'(gng^{-1})^{-1} \in H_p' \qquad (15.93)$$

for all $n \in N(H_p)$. This means,

$$gN(H_p)g^{-1} \subset N(H_p'). \qquad (15.94)$$

If follows that since conjugation preserves isomorphism,

$$|gN(H_p)g^{-1}| = |N(H_p')|. \qquad (15.95)$$

Equations (15.94) and (15.95) together imply,

$$gN(H_p)g^{-1} = N(H'_p) \qquad (15.96)$$

This proves that for any given p, normalizers of the sylow p subgroups are conjugate to each other. $\boxed{\text{QED}\checkmark}$

Proof of Theorem By sylow theory we are guaranteed the existence of subgroups of order 5. Such that,

$$n_5(G) \simeq 1 (mod\ 5) \qquad (15.97)$$

and

$$n_5(G) \mid 2^2 \cdot 3 \qquad (15.98)$$

The candidates include:

$$n_5(G) \in \{1,\ 6\} \qquad (15.99)$$

However, by Lemma (15.14) above which states that $n_p(G) > 1$ if G is simple, we know,

$$n_5(G) \neq 1. \qquad (15.100)$$

Therefore

$$n_5(G) = 6. \qquad (15.101)$$

Since 5 is a prime, $\mathbb{Z}/5$ is the only isomorphism class for groups of order 5. Therefore all six Sylow 5 subgroups are isomorphic to $\mathbb{Z}/5$, i.e.

$$H_5 \simeq \mathbb{Z}/5 \qquad (15.102)$$

for all sylow 5 subgroups. Given a generator x of a sylow 5 subgroup, i.e. $H_5 = \langle x \rangle$. The elements x^2, x^3, and x^4 are also generators of H_5. Hence each sylow 5 subgroup has four generators. Furthermore, any distinct pair of sylow 5 subgroups intersect only in the identity, i.e.

$$H_5 \cap H_5' = \{e\} \text{ if } H_5 \neq H_5'. \qquad (15.103)$$

If by contradiction we assume some there exists some non-trivial $y \in H_5 \cap H_5'$, then by Lagrange, $|y|$ divides $5 = |H_5| = |H_5'|$. Then since 5 is a prime, it must follow that $|y| = 5$ also. This means $H_5 = \langle y \rangle$ and $H_5' = \langle y \rangle$. And this is a contradiction.

This implies that the number of elements of order 5 in G is,

$$n_5(G) \cdot (5 - 1) = 6 \cdot 4 = 24 \qquad (15.104)$$

This accounts for all order 5 elements in G. Next, to determine the order of the normalizer of H_5, we invoke the counting formula,

$$|\Theta_H| = \frac{|G|}{|Stab_G(H)|}. \qquad (15.105)$$

One recognizes that under the group action of conjugation on the set of sylow 5 subgroups, the orbit Θ_H is the set of all sylow 5 subgroups, and the stabilizer is the normalizer of any sylow 5 subgroup. This yields,

$$n_5(G) = \frac{|G|}{|N(H_5)|}. \qquad (15.106)$$

Plugging in the appropriate values we get,

$$6 = \frac{60}{|N(H_5)|}, \qquad (15.107)$$

which implies

$$|N(H_5)| = 10 \qquad (15.108)$$

Lemma 15.16. *There is no subgroup of order 15 in G*

Proof of Lemma: By way of contradiction, assume that there exists a subgroup H_{15} of order 15 in G. Then because H_{15} is itself a group, the Sylow theorems apply to it.

$$5 \mid H \qquad (15.109)$$

therefore there exists some sylow 5 subgroup $\tilde{H}_5$ inside H_{15}. According to the sylow theorems the number $\tilde{n}_5$ of such sylow 5 subgroups must be such that,

$$\tilde{n}_5 \mid 3 \qquad (15.110)$$

and

$$\tilde{n}_5 \equiv 1 \ (mod \ 5). \qquad (15.111)$$

The only permissible value is

$$\tilde{n}_5 = 1. \qquad (15.112)$$

This implies that $\tilde{H}_5$ is characteristically normal in H_{15},

$$\tilde{H}_5 \lhd H_{15}. \qquad (15.113)$$

Hence,

$$x\tilde{H}_5 x^{-1} \subset \tilde{H}_5 \qquad (15.114)$$

for all $x \in H_{15}$. However, by sylow theorem and even by the Cauchy theorem, there exists some $x_3 \in H_{15}$, such that $|x_3| = 3$. By the above statement,

$$x_3 \tilde{H}_5 x_3^{-1} \subset \tilde{H}_5. \qquad (15.115)$$

This implies that

$$x_3 \in N(\tilde{H}_5). \qquad (15.116)$$

It follows from Lagrange's theorem that,

$$|x_3| \mid |N(\tilde{H}_5)|. \qquad (15.117)$$

However from Equation (15.108), $|N(\tilde{H}_5)| = 10$, therefore,

$$3 \mid 10. \tag{15.118}$$

This is clearly a contradiction. Therefore if G is a simple group of order 60, it can have no subgroups of order 15.
QED✓

Corollary 15.17. *If G is a simple group of order* 60, *and H a subgroup of G, then $[G : H]$ can never equal* 4. *In particular, the number of sylow p subgroups, $n_p(G)$, can never equal* 4.

Proof of Corollary: By Lemma (15.16),

$$|H| \neq 15. \tag{15.119}$$

$$[G : H] = \frac{|G|}{|H|} \tag{15.120}$$

$$|H| = \frac{|G|}{[G : H]} \tag{15.121}$$

By way of contradiction, assume $[G : H] = 4$ for some H. Then,

$$|H| = 60/4 = 15 \tag{15.122}$$

This is a contradiction, and proves the first part of the corollary. QED✓

For the proof of the second part we recall the following. The group action of conjugation on the set of sylow p subgroups is a transitive action hence it consists of a single orbit. The size of this orbit is the set of sylow p subgroups, $n_p(G)$. And the stabilizer of any of the sylow p subgroups is the normalizer of the subgroup. In this context, the orbit-stabilizer theorem is therefore written as,

$$n_p(G) = \frac{|G|}{|N(H_p)|} = [G : N(H_p)], \qquad (15.123)$$

and therefore from the first part of the corollary, it follows that

$$n_p(G) \neq 4. \qquad (15.124)$$

QED✓

We next proceed to consider the elements of G which have order 3, i.e,

$$x \in G \text{ such that } |x| = 3. \qquad (15.125)$$

By the sylow theorems, there exists some sylow 3 subgroup H_3 in G. And the number $n_3(G)$ of sylow 3 subgroups is such that

$$n_3(G) \mid (2^2 \cdot 5 = 20) \qquad (15.126)$$

and

$$n_3(G) \equiv 1 \ (mod \ 3). \qquad (15.127)$$

The candidates satisfying the above two sylow conditions are given by,

$$n_3(G) \in \{1, \ 4, \ \text{and,} \ 10\}. \qquad (15.128)$$

However by Lemma (15.14),

$$n_3(G) \neq 1, \qquad (15.129)$$

and by the corollary to Lemma (15.16),

$$n_3(G) \neq 4. \qquad (15.130)$$

Therefore it must be the case that,

$$n_3(G) = 10. \qquad (15.131)$$

Then since 3 is a prime, the sylow 3 subgroups are all isomorphic to $\mathbb{Z}/3$, i.e.

$$H_3 \simeq \mathbb{Z}/3. \tag{15.132}$$

The Euler totient function for 3 is

$$\phi(3) = 2, \tag{15.133}$$

therefore there are two generators of each Sylow 3 subgroup. And according to the same reasoning outlined in the case of the sylow 5 subgroups above, the intersection of any pair of distinct sylow 3 subgroups can only contain the identity, i.e.,

$$H_3 \cap H_3' = \{e\} \text{ for all } H_3 \neq H_3'. \tag{15.134}$$

Therefore the total number of elements of order 3 in G is given by,

$$|\{x \in G \text{ such that } |x| = 3\}| = n_3(G) \cdot (3 - 1) \tag{15.135}$$
$$= 10 \cdot 2$$
$$= 20.$$

So this far in the counting process we have counted all elements of order one of which there is only one, the identity; we have counted all elements of order 5, of which we determined there are 24; and we have counted all elements of order 3, of which we just determined there are 20. So far we have accounted for,

$$|\{e\}| + |\{x, |x| = 5\}| + |\{x, |x| = 3\}| = \tag{15.136}$$
$$1 + 24 + 20 = 45.$$

This leaves,

$$|G| - 45 = \tag{15.137}$$
$$60 - 45 = 15$$

elements to be accounted for. At this point in the proof, we have built-in enough implicit constraints to proceed. We will next use these constraints to determine what type(s) of elements occupy the remaining 15 spaces.

Lemma 15.18. *The normalizers of the sylow 5 subgroups are isomorphic to* D_{10}.

Proof of Lemma: Recall from Lemma (15.15) that for a given p, all the sylow p subgroup normalizers are isomorphic to each other. Also recall that there are only two isomorphism classes for groups of order 10. They are either isomorphic to D_{10} or to $\mathbb{Z}/10$. By way of contradiction, let us assume that normalizers of the sylow 5 subgroups are isomorphic to $\mathbb{Z}/10$. Note that the intersection of the normalizers of any two distinct sylow 5 subgroups cannot contain any generators of $\mathbb{Z}/10$. In other words the intersection cannot contain any elements of order 10. Because if they do, then they are the same subgroup and the respective sylow 5 subgroups are also the same. This follows from the definition of normalizer and from the second sylow theorem–i.e the conjugacy of the sylow p subgroups. Therefore if

$$x \in N(H_5) \cap N(\tilde{H}_5) \tag{15.138}$$

where

$$H_5 \neq \tilde{H}_5, \tag{15.139}$$

then

$$|x| \neq 10. \tag{15.140}$$

The number of generators of $\mathbb{Z}/10$ is given by the Euler totient function,

$$\phi(10) = |\{1, 3, 7, 9\}| = 4. \tag{15.141}$$

Hence if $N(H_5) \simeq \mathbb{Z}/10$, then each sylow 5 subgroup is associated with four distinct elements of order 10. This will contribute $4 \cdot n_5(G)$ number of elements of order 10 to our running tally of elements of G,

$$4 \cdot n_5(G) = 4 \cdot 6 = 24 \tag{15.142}$$

This will bring our tally to,

$$1 + 24 + 20 + 24 = 69 > 60. \tag{15.143}$$

This is a contradiction to the order of G which we know to be 60. This concludes our proof of Lemma (15.18), i.e. that the normalizers of the sylow 5 subgroups are isomorphic to D_{10} and not to $\mathbb{Z}/10$. $\boxed{\text{QED}\checkmark}$

Corollary 15.19. *If G is a simple group of order 60, then it has no subgroups isomorphic to $\mathbb{Z}/10$*

Proof of Corollary: By way of contradiction, assume there exists a subgroup H_z of G such that $H_z \simeq \mathbb{Z}/10$. Then since the subgroup H_z is itself a group, the sylow theorems apply to it. Therefore there exists a subgroup, $H_{z,5}$, of order 5 contained in H_z. Since $H_z \simeq \mathbb{Z}/10$ is abelian, every of its subgroups are normal in H_z. In particular, $H_{z,5}$ is normal in H_z. Hence H_z is in the normalizer of $H_{z,5}$. i.e.,

$$H_z = \mathbb{Z}/10 \subset N(H_{z,5}). \tag{15.144}$$

However according to Lemma (15.18) the normalizers of the sylow 5 subgroups are isomorphic to D_{10} and not to $\mathbb{Z}/10$. Therefore,

$$H_z \simeq \mathbb{Z}/10 \subset N(H_{z,5}) \simeq D_{10} \tag{15.145}$$

implying

$$\mathbb{Z}/10 \simeq D_{10}. \qquad (15.146)$$

This is clearly a contradiction. It concludes our proof that any simple group of order 60 can have no subgroups isomorphic to $\mathbb{Z}/10$. $\boxed{\text{QED}\checkmark}$

Lemma 15.20. *The normalizers of the sylow 3 subgroups are isomorphic to D_6.*

Proof of Lemma: Recall that there are only two isomorphism classes for groups of order 6. They are either isomorphic to $\mathbb{Z}/6$ or to D_6, the dihedral group of order 6. Assume by way of contradiction that there existed some simple group of order 60 for which the normalizers of the sylow 3 subgroups were isomorphic to $\mathbb{Z}/6$. Note that the intersection of any two distinct such normalizers cannot contain any generators. i.e. if

$$x \in N(H_3) \cap N(\tilde{H}_3) \qquad (15.147)$$

where

$$H_3 \neq \tilde{H}_3, \qquad (15.148)$$

then

$$|x| \neq 6. \qquad (15.149)$$

The number of generators of $\mathbb{Z}/6$ is given by the Euler totient function,

$$\phi(6) = |\{1, 5\}| = 2. \qquad (15.150)$$

Hence if $N(H_3) \simeq \mathbb{Z}/6$, then each sylow 3 subgroup is associated with 2 distinct elements of order 6. This will contribute $2 \cdot n_3(G)$ number of elements of order 6 to our running tally of elements of G,

$$2 \cdot n_3(G) = 2 \cdot 10 = 20 \qquad (15.151)$$

This will bring our tally to,

$$1 + 24 + 20 + 20 = 65 > 60. \qquad (15.152)$$

This is a contradiction to the order of G which we know to be 60. This concludes our proof of Lemma (15.20), i.e. that the normalizers of the sylow 3 subgroups are isomorphic to D_6 and not to $\mathbb{Z}/6$. $\boxed{\text{QED}\checkmark}$

Corollary 15.21. *If G is a simple group of order 60, then it has no subgroups isomorphic to $\mathbb{Z}/6$.*

Proof of Corollary: Here, we proceed as we did in the above proof of Corollary (15.19) that G has no subgroups isomorphic to $\mathbb{Z}/10$. Again by way of contradiction, assume there exists a subgroup H_z of G such that $H_z \simeq \mathbb{Z}/6$. Then since the subgroup H_z is itself a group, the sylow theorems apply to it. Therefore there exists a subgroup, $H_{z,3}$, of order 3 contained in H_z. Since $H_z \simeq \mathbb{Z}/6$ is abelian, every of its subgroups are normal in H_z. In particular, $H_{z,3}$ is normal in H_z. Hence H_z is in the normalizer of $H_{z,3}$. i.e.,

$$H_z = \mathbb{Z}/6 \subset N(H_{z,3}) \qquad (15.153)$$

However according to Lemma (15.20) the normalizers of the sylow 3 subgroups are isomorphic to D_6 and not to $\mathbb{Z}/6$. Therefore,

$$H_z \simeq \mathbb{Z}/6 \subset N(H_{z,3}) \simeq D_6 \qquad (15.154)$$

implying

$$\mathbb{Z}/6 \simeq D_6. \qquad (15.155)$$

This is clearly a contradiction. It concludes our proof that any simple group of order 60 can have no subgroups isomorphic to $\mathbb{Z}/6$. $\boxed{\text{QED}\checkmark}$

Now, fully equipped with the above Lemmas and with our tally thus far, we return to the Euler product of G and the Sylow theorems.

$$|G| = 2^2 \cdot 3 \cdot 5. \tag{15.156}$$

According to our tally we are yet to account for 15 elements. We have accounted for the identity, all elements of order 3, and all elements of order 5. According to Lagrange's theorem, the remaining elements must either have order 2 or order 4. According to the Sylow theorems, G contains a subgroup of order $2^2 = 4$. Recall that for groups of order 4 there are only two isomorphism classes. They are either isomorphic to $\mathbb{Z}/4$ or to the Klein four group, $\mathbb{Z}/2 \times \mathbb{Z}/2$. In either case, they contain a subgroup of order 2. Hence our group G contains at least one element, x_0, of order 2. Since conjugacy preserves order of elements, one reasonable next step is to count all elements in C_{x_0}, the conjugacy class of x_0. To do this, all we require is $|C(x_0)|$, the order of the center of x_0, with which we can solve the equation,

$$|C_{x_0}| = \frac{|G|}{|C(x_0)|}. \tag{15.157}$$

Furthermore, we know that because $x_0 \in \mathbb{Z}/4$ or $x_0 \in \mathbb{Z}/2 \times \mathbb{Z}/2$, which are both abelian groups. Therefore,

$$\mathbb{Z}/4 \subset C(x_0) \tag{15.158}$$

or

$$\mathbb{Z}/2 \times \mathbb{Z}/2 \subset C(x_0), \tag{15.159}$$

whichever is applicable. In the first case we have,

$$|\mathbb{Z}/4| \leq |C(x_0)|, \tag{15.160}$$

and in the second case we have

$$|\mathbb{Z}/2 \times \mathbb{Z}/2| \leq |C(x_0)|. \tag{15.161}$$

Whatever the case, it holds that,

$$4 \leq |C(x_0)|. \tag{15.162}$$

Now we note that since $|G| = 4 \cdot 3 \cdot 5$ and by Lagrange $|C(x_0)|$ divides $|G|$, the only possible values of $|C(x_0)|$ which are greater than 4 must be divisible by either 3 or 5. Specifically, such a number must be in the set $\{6, 10, 12, 15, 20, 30, 60\}$. If indeed $|C(x_0)| > 4$, then since

$$3 \mid |C(x_0)| \text{ or } 5 \mid |C(x_0)|, \tag{15.163}$$

Case 1. If 3 divides $|C(x_0)|$ then according to Cauchy's theorem there exists $y_3 \in C(x_0)$ such that $|y_3| = 3$. Then the element

$$y_3 x_0 \in C(x_0) \tag{15.164}$$

and

$$|y_3 x_0| = 3 \cdot 2 = 6. \tag{15.165}$$

And

$$\langle y_3 x_0 \rangle \simeq \mathbb{Z}/6 \subset G. \tag{15.166}$$

This is a contradiction, since according to Corollary (15.21), G contains no subgroups isomorphic to $\mathbb{Z}/6$.

Case 2. If 5 divides $|C(x_0)|$ then according to Cauchy's theorem there exists $y_5 \in C(x_0)$ such that $|y_5| = 5$. Then the element

$$y_5 x_0 \in C(x_0) \tag{15.167}$$

and

$$|y_5 x_0| = 5 \cdot 2 = 10. \tag{15.168}$$

And

$$\langle y_5 x_0 \rangle \simeq \mathbb{Z}/10 \subset G. \qquad (15.169)$$

This also is a contradiction, since according to Corollary (15.19), G contains no subgroups isomorphic to $\mathbb{Z}/10$. It therefore holds that,

$$|C(x_0)| = 4. \qquad (15.170)$$

This implies that,

$$|C(x_0)| = \frac{|60|}{4} = 15 \qquad (15.171)$$

i.e. the conjugacy class of x_0 contains all 15 elements we were seeking. And as noted, since conjugacy preserves order, they are all elements of order 2. Since there are no elements of order 4, it follows that all the sylow 2 subgroups in G are isomorphic to the Klein group, $\mathbb{Z}/2 \times \mathbb{Z}/2$.

Proposition 15.22. *The number of sylow 2 subgroups is 5, i.e.* $n_2(G) = 5$.

Proof of Proposition: We showed above that any sylow 2 subgroup, H_2, is isomorphic to the Klein four group, $\mathbb{Z}/2 \times \mathbb{Z}/2$. We also showed that for any element x of order 2 in G,

$$x \in \mathbb{Z}/2 \times \mathbb{Z}/2 \simeq H_2, \qquad (15.172)$$

and the order of the center of x is 4, i.e $|C(x)| = 4$. Since the Klein four group is abelian it follows that,

$$C(x) = H_2. \qquad (15.173)$$

Therefore given any two distinct sylow 2 subgroups H_2 and H_2', if there exists an $x_0 \in H_2 \cap H_2'$, then since such x_0 commutes with every element of H_2 as well as every element of H_2', it follows that $C(x_0)$ contains all elements in H_2 and all elements in H_2'. However since $H_2 \neq H_2'$ there is at least one element in H_2 not in H_2'. Hence the sum of distinct elements in H_2 and H_2' is at least $4+1 = 5$. This exceeds the number of elements in $C(x_0)$ and is therefore a contradiction. So indeed the intersection of any two sylow 2 subgroups contains only the identity.

$$H_2 \cap H_2' = \{e\} \tag{15.174}$$

for any $H_2 \neq H_2'$. Therefore distributing the 15 elements of order 2 in G into distinct Klein four groups requires that each group exclusively contains three non-trivial elements. This implies there are exactly 5 sylow-2 subgroups in G. $\boxed{\text{QED}\checkmark}$

Applying this to the counting formula,

$$n_2(G) = \frac{|G|}{|N(H_2)|}, \tag{15.175}$$

we get,

$$|N(H_2)| = \frac{60}{5} = 12. \tag{15.176}$$

With this we form the order 5 coset space, $G/N(H_2)$. We next invoke the homomorphic map,

$$\Pi : G \to Bij(G/N(H_2), G/N(H_2)) \tag{15.177}$$

defined by the left regular action of G on the coset space as follows,

$$\Pi(g) = \pi(g) \tag{15.178}$$

where,

$$\pi(g)(kN(H_2)) = gkN(H_2). \tag{15.179}$$

In other words, Π is a mapping of G into the symmetric group on $|G/N(H_2)|$ letters. Therefore,

$$\Pi : G \to S_5. \tag{15.180}$$

Let $\tilde{\Pi}$ be a mapping given by,

$$\tilde{\Pi} : G \to Image(G) \subset S_5, \qquad (15.181)$$

such that

$$\tilde{\Pi}(g) = \Pi(g). \qquad (15.182)$$

Then since $\tilde{\Pi}$ is a homomorphism between G and $Image(G) \subset S_5$, if $\tilde{\Pi}$ is one-to-one, then $\tilde{\Pi}$ is an isomorphism and therefore G is isomorphic to a subgroup of S_5. Recall that a homomorphic map is one-to-one if and only if its kernel is trivial. By the first isomorphism theorem, the kernel of any homomorphic map from G is normal in G.

$$Kern\left(\tilde{\Pi}\right) \triangleleft G. \qquad (15.183)$$

However, G is simple and therefore its only normal subgroups are $\{e\}$ and G itself. To check if $Kern\left(\tilde{\Pi}\right) = G$, we note that $Kern\left(\tilde{\Pi}\right)$ is the set of elements in G which do not induce a non-trivial permutation on the coset space $G/N(H_2)$. Therefore,

$$Kern\left(\tilde{\Pi}\right) \subset N(H_2) \subsetneq G. \qquad (15.184)$$

And Hence,

$$Kern\left(\tilde{\Pi}\right) \neq G. \qquad (15.185)$$

Therefore,

$$Kern\left(\tilde{\Pi}\right) = \{e\}. \qquad (15.186)$$

It follows that G is isomorphic to a proper subgroup of S_5.

Lemma 15.23. *Any subgroup of order* 60 *in* S_5 *is isomorphic to* A_5.

Proof of Lemma: Let G be an arbitrary subgroup of order 60 in S_5. By the index 2 theorem,

$$G \lhd S_5 \tag{15.187}$$

and

$$A_5 \lhd S_5. \tag{15.188}$$

It follows that,

$$A_5 \cap G \lhd A_5. \tag{15.189}$$

Recall that A_5 is simple. Therefore every normal subgroup it contains is either A_5 (i.e. $A_5 \cap G = A_5$, hence $G = A_5$ and we are done with the proof) or $A_5 \cap G = \{e\}$ (in which case $G \neq A_5$, and our proposition was false). By way of contradiction, assume $A_5 \cap G = \{e\}$. i.e. Other than $\{e\}$ which is shared by A_n and G, they each contain 59 elements which they do not share. Then,

$$S_5 \setminus (A_5 \cup G) = \{\tau\}, \tag{15.190}$$

where $\{\tau\}$ is the sole element of S_5 not contained in either A_5 or G. It follows that since by definition A_5 contains all (and only) even permutations, G must contain only odd permutations. Furthermore it must contain at least all but one odd permutations – the only odd permutation it excludes is $\{\tau\}$. Therefore G contains at least two disjoint 2-cycles. Specifically, given any 2-cycle, σ there are 3 other 2-cycles in S_5 from which it is disjoint. If $\{\tau\}$ is one of the 2-cycles from which it is disjoint, this still leaves two other options in G, call them ρ and π. Then the product $\sigma\rho$ and $\sigma\pi$ are both in G and are even permutations, hence are non-trivial constituents of the intersection of A_5 and G. This is a contradiction. Hence, $G \simeq A_5$. $\boxed{\text{QED}\checkmark}$

This completes the overall proof of our theorem that every simple group of order 60 is isomorphic to A_5. $\boxed{\text{QED}\checkmark}$

15.4   Exercises

1. List all isomorphism classes for groups of order 67.

2. List all isomorphism classes for groups of order 34.

3. List all isomorphism classes for groups of order 15.

4. List all isomorphism classes for groups of order 51.

5. List all isomorphism classes of groups of order 781

6. Show that A_4 is not simple.

7. Prove that all groups of order pq where p and q are primes and $p < q$ have a normal sylow q subgroup.

8. Let G be a group of order pq where p, q are primes, $p < q$, and $p \nmid q - 1$. Prove that the sylow p subgroup is normal in G.

9. Let G be a group of order pq where p, q are primes, $p < q$, and $q \not\equiv 1 (mod\ p)$. Prove that $G \simeq \mathbb{Z}/pq$.

10. Let G be a group whose order is the product of powers of two distinct primes p and q, such that the number of sylow p and sylow q subgroups each equals one. Prove that the elements of the sylow p and sylow q subgroups commute with each other.

11. Prove that G equals the direct product of its sylow subgroups if and only it each of its sylow subgroups are normal in G.

12. Let G be group of order pq where p, q are primes and $p < q$. Under what condition is G isomorphic to the direct product of its sylow p and sylow q subgroups?

13. Prove that any group of order 45 is solvable.

14. Prove that any group of order 539 is solvable.

15. Prove that every simple group of order 60 is isomorphic to A_5.

16. Let G be a group with $|G| = p^2q$, where p, q are primes and $p < q$, show that G is solvable.

17. Let G be a group with $|G| = p^2q$ where p, q are primes and $p < q$. If $q \not\equiv 1(mod\ p)$ and $p^2 \not\equiv 1(mod\ q)$, show that G is abelian.

18. Prove that every simple group of order 60 is isomorphic to A_5

19. Prove that there are 13 isomorphism classes for groups of order 60.

Chapter 16

Finite Abelian Groups

As we close in on the notion and outline of classification of finite groups into isomorphism classes, we must mark a certain critical milestone: the classification of finite abelian groups. The tractability of abelian groups relative to their non-abelian counterparts is a key feature that permeates all of algebra. We will see in this chapter that finite abelian groups can always be expressed as direct products of cyclic groups of prime power order. This is called the fundamental theorem of finite abelian groups, and it is fundamental indeed. It confers remarkable simplicity and insight into the behavior and structure of abelian groups. At the end of this chapter we should be able to list all the isomorphism classes of abelian groups of a given order.

16.1 Fundamental Thm. of Finite Abelian Groups

Theorem 16.1. *Every finite abelian group is isomorphic to the direct product of a unique set of cyclic groups of prime power order.*

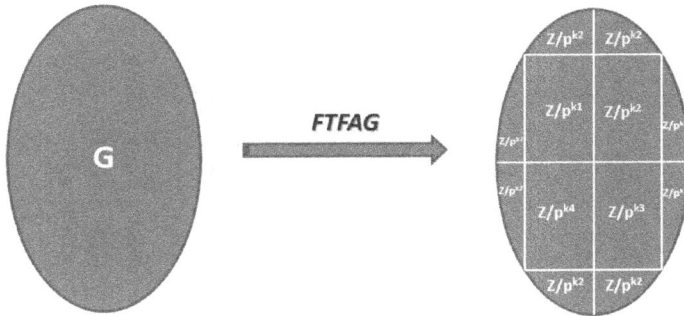

Figure 16.1: Fundamental Theorem of Finite Abelian Groups

Proof: $|G| = \prod_{i=1}^{n} p_i^{k_i}$. By sylow theorems, for each i, there exists a sylow subgroup, i.e. a subgroup H_i such that $|H_i| = p_i^{k_i}$. By the abelian-ness of G, each of these sylow subgroups are normal in G. Therefore by Theorem (15.7), G is a direct product of each of these sylow p-subgroups. The collection is unique at this level because of the normality in G of each of the sylow p-subgroups. It remains to show that each of these sylow p subgroups are themselves isomorphic to the direct product of a unique set of cyclic groups. We restate our task in the following lemma:

Lemma 16.2. *If G is an abelian group of prime power order, then G is isomorphic to the direct product of a unique collection of cyclic groups.*

Proof of lemma:

Claim: Let G be an abelian group of prime power order, and $x \in G$ be an element of maximal order in G. And let H be the largest subgroup of G for which $\langle x \rangle \cap H = \{e\}$. Then it follows that $G \simeq \langle x \rangle \times H$.

By "*x having maximal order in G*" we mean that given $|x| = p^r$, it follows that $|g| \leq p^r$ for all $g \in G$. Also note that by the maximality condition on H, if $\langle x \rangle \cap U = \{e\}$ then $|U| \leq |H|$.

Proof of claim: From the proof of Theorem (15.7) and the section on internal direct products we note that since $\langle x \rangle \cap H = \{e\}$, and both $\langle x \rangle$ and H are normal in G, all that remains is to show that $\langle x \rangle H = G$. By way of contradiction, let us assume there exists an element $y \in G$ such that $y \notin \langle x \rangle H$. Note that we can always choose a y such that $y^{p-1} \notin \langle x \rangle H$, but $y^p \in \langle x \rangle H$. To see this, note that since $y^{p^r} = e \in \langle x \rangle H$, if we successively raise y to powers of p, we eventually end-up in $\langle x \rangle H$. We particularly pick the minimum power to which we can raise y to get into $\langle x \rangle H$. i.e. we pick the power $w \in \mathbb{N}$ such that $y^{p^w} \in \langle x \rangle H$ and $y^{p^{w-1}} \notin \langle x \rangle H$. We then simply choose via substituting $y^{p^{w-1}} \to y$ and therefore $y^{p^w} \to y^p$.

Now since $y^p \in \langle x \rangle H$, it follows that $y^p x^{-a} = h$ for some integer $a \leq p^r$ and some $h \in H$. Therefore, $e = y^{p^r} = (y^p)^{p^{r-1}} = (x^a h)^{p^{r-1}} = x^{ap^{r-1}} h^{p^{r-1}}$. Therefore $x^{ap^{r-1}}$ is the inverse of $h^{p^{r-1}}$. Hence $x^{ap^{r-1}}$ is in H and clearly also in $\langle x \rangle$. Therefore $x^{ap^{r-1}} \in H \cap \langle x \rangle$, hence $x^{ap^{r-1}} = e$. Therefore, $p^r \mid ap^{r-1}$. This in turn implies that $p \mid a$. Hence $a = pm$ for some $m \in \mathbb{N}$. Therefore $y^p = x^{pm} h$. Let $U = \langle yx^{-m} \rangle H$. Note that because $y \notin \langle x \rangle H$ it follows that $yx^{-m} \notin H$. Therefore H is a proper subgroup of U. Hence by the maximality condition of H, it follows that there exists some non-trivial $v \in \langle x \rangle \cap U$. Therefore,

$$v = x^l = (yx^{-m})^s \tilde{h} \tag{16.1}$$

for some natural numbers $l, s \leq p^r$ and some $\tilde{h} \in H$.

We claim that $p \nmid s$. To see this, assume by contradiction that $p \mid s$, then $s = pt$ for some $t \in \mathbb{N}$. Therefore $v = x^l = (yx^{-m})^{pt}\tilde{h} = (y^p x^{-mp})^t \tilde{h} = h^t \tilde{h} \in H$. Hence $v \in H \cap \langle x \rangle$, and therefore $v = e$. This is a contradiction.

Now since $p \nmid s$, by bezout's identity it follows that $1 = pi + sj$ for some $i, j \in \mathbb{Z}$. It follows that $y = y^1 = y^{pi+sj} = (y^p)^i (y^s)^j$. From the equation for v, we can express y^s as,

$$y^s = x^l x^{ms} \tilde{h}^{-1} \in \langle x \rangle H \qquad (16.2)$$

Therefore $(y^s)^j \in \langle x \rangle H$. In addition, we have from earlier that $y^p \in \langle x \rangle H$, and therefore $(y^p)^i \in \langle x \rangle H$. Putting it together, we see that $y = (y^p)^i (y^s)^j \in \langle x \rangle H$. This is a contradiction because we carefully selected y to guarantee that $y \notin \langle x \rangle H$. This completes the proof of the fundamental theorem of finite abelian groups. $\boxed{\text{QED}\checkmark}$.

Example 16.3. Given $G \simeq \mathbb{Z}/5 \times \mathbb{Z}/5 \times \mathbb{Z}/7$ we say that the *elementary divisors* of G are the numbers 5, 5, and 7. In other words, the elementary divisors of G are the orders of the prime powered cyclic subgroup factors of G.

16.2 Invariant Factor Decomposition

We saw in the fundamental theorem of finite abelian groups that any finite abelian group G can be decomposed into a unique collection of factors of cyclic groups of prime power order. For instance,

Example 16.4. $G = \mathbb{Z}/5 \times \mathbb{Z}/5 \times \mathbb{Z}/7 \simeq \mathbb{Z}/5 \times \mathbb{Z}/35$

The isomorphism above is because 5 and 7 are relatively prime. In the above example, we say the *elementary divisors* of G are 5, 5 and 7 and the *invariant factors* of G are 5 and 35. More generally, it holds that any finite abelian group has an invariant factor decomposition of the form,

$$G = \mathbb{Z}/b_1 \times \mathbb{Z}/b_2 \times \cdots \times \mathbb{Z}/b_n, \qquad (16.3)$$

such that

$$b_1 \mid b_2 \mid \cdots \mid b_n$$

just as $5 \mid 35$ in the example. The set $\{b_1 \cdots b_n\}$ are the invariant factors of G.

Table 16.1: Finite Abelian Group Factorization

Elementary Decomposition	Invariant Factor Decomposition
$\mathbb{Z}/5 \times \mathbb{Z}/5 \times \mathbb{Z}/7$	$\mathbb{Z}/5 \times \mathbb{Z}/35$
$\mathbb{Z}/2 \times \mathbb{Z}/2 \times \mathbb{Z}/2 \times \mathbb{Z}/5 \times \mathbb{Z}/7$	$\mathbb{Z}/2 \times \mathbb{Z}/140$
$\mathbb{Z}/2 \times \mathbb{Z}/2 \times \mathbb{Z}/2 \times \mathbb{Z}/3 \times \mathbb{Z}/3 \times \mathbb{Z}/5$	$\mathbb{Z}/2 \times \mathbb{Z}/6 \times \mathbb{Z}/30$
$\mathbb{Z}/2 \times \mathbb{Z}/3 \times \mathbb{Z}/3 \times \mathbb{Z}/5$	$\mathbb{Z}/3 \times \mathbb{Z}/30$
$\mathbb{Z}/3 \times \mathbb{Z}/5 \times \mathbb{Z}/7$	$\mathbb{Z}/105$

The invariant factor decomposition is arrived at by arranging the factors row-wise and then multiplying column-wise. Table (16.2) is an example of how we obtained $\mathbb{Z}/2 \times \mathbb{Z}/2 \times \mathbb{Z}/2 \times \mathbb{Z}/3 \times \mathbb{Z}/3 \times \mathbb{Z}/5 \simeq \mathbb{Z}/2 \times \mathbb{Z}/6 \times \mathbb{Z}/30$. And Table (16.3) is an example of how we obtained $\mathbb{Z}/2 \times \mathbb{Z}/3 \times \mathbb{Z}/3 \times \mathbb{Z}/5 \simeq \mathbb{Z}/3 \times \mathbb{Z}/30$

Table 16.2: Invariant Factorization Example

$$
\begin{array}{ccc}
2 & 2 & 2 \\
  & 3 & 3 \\
\times & & 5 \\
\hline
2 & 6 & 30
\end{array}
$$

Table 16.3: Invariant Factorization Example

$$
\begin{array}{cc}
 & 2 \\
3 & 3 \\
\times & 5 \\
\hline
3 & 30
\end{array}
$$

Example 16.5. Find the elementary divisors and the invariant factors of an abelian group G of order 90.

The prime factorization of $90 = 2 \cdot 3^2 \cdot 5$. There is only one factor of order 2 and only one of order 5, hence both 2 and 5 must be included in the set of elementary divisors of G. There are two factors of order 3, hence two possibilities: $3 \cdot 3$ or 9. In the first case, the elementary factorization is $G \simeq \mathbb{Z}/2 \times \mathbb{Z}/3 \times \mathbb{Z}/3 \times \mathbb{Z}/5$ and the invariant factorization is $G \simeq \mathbb{Z}/3 \times \mathbb{Z}/30$, hence the elementary divisors are $\{2, 3, 3, 5\}$ and the invariant factors are $\{3, 30\}$. In the second case, the elementary factorization is $G \simeq \mathbb{Z}/2 \times \mathbb{Z}/9 \times \mathbb{Z}/5$ and the invariant factorization is $G \simeq \mathbb{Z}/90$, hence the elementary divisors are $\{2, 9, 5\}$ and the invariant factor is 90.

16.3   Exercises

1. Find all abelian groups of order 180 up to isomorphism

2. Find all abelian groups of order 540 up to isomorphism

3. Find the elementary factor decompositions and the corresponding invariant factor decompositions of

$$G = \mathbb{Z}/2 \times \mathbb{Z}/12 \times \mathbb{Z}/150$$

4. Find the elementary factor decompositions and the corresponding invariant factor decompositions of

$$G = \mathbb{Z}/4 \times \mathbb{Z}/18 \times \mathbb{Z}/60$$

5. Let G be a group of order 40 such that no element has order exceeding 20. Find all admissible elementary factor decompositions of G.

6. Find all abelian groups of order 9900 up to isomorphism

7. List all abelian groups of order 29,400 up to isomorphism

8. Given two finite abelian groups U and V such that $U \times U \simeq V \times V$, prove that $U \simeq V$

9. Classify up to isomorphism, all abelian groups of order pq where p and q are distinct primes.

10. Classify up to isomorphism, all abelian groups of order pqr where p, q, and r are distinct primes.

11. Describe all finite abelian simple groups.

12. Prove that if G is an abelian group of odd order, then $\prod_{g \in G} g = e$.

13. If G is an abelian group of even order then what does $\prod_{g \in G} g$ yield?

14. G is an abelian group of order 168, determine the isomorphism class of G that has

    a) exactly three elements of order 2.

    b) exactly 2 elements of order 2.

15. Show that if $G/Z(G)$ is cyclic, then G is abelian.

Chapter 17

Finite Group Classification Overview

The isomorphism classification of finite groups presents a way of studying groups up to isomorphism, hence distilling their salient structure which is independent of representation. Here we show a table of the isomorphism classes of groups of small order, in particular, groups with order ≤ 15. We also nod towards the classification of finite simple group. That concerted effort of collaborative mathematics spanned decades, involved several groups of prominent mathematicians, and introduced some single proofs which were additionally notable for their enormous lengths, sometimes exceeding one thousand pages. Needless to say, a full delineation of that work is beyond the aims of this book. For our purposes, it is sufficient to recognize that at the heart of that entire exercise are the sylow theorems and the counting methods and approaches with which we proved them.

17.1 Groups of Small Order

Table 17.1: Groups of Small Order

$|G|$	Group	Name or Type	Abelian
1	$\{e\}$	Identity	Yes
2	$\mathbb{Z}/2$	Cyclic	Yes
3	$\mathbb{Z}/3$	Cyclic	Yes
4	$\mathbb{Z}/4$	Cyclic	Yes
	$\mathbb{Z}/2 \times \mathbb{Z}/2$	Klein four	Yes
5	$\mathbb{Z}/5$	Cyclic	Yes
6	$\mathbb{Z}/6 \simeq \mathbb{Z}/3 \times \mathbb{Z}/2$	Cyclic	Yes
	S_3	Symmetric	No
7	$\mathbb{Z}/7$	Cyclic	Yes
8	$\mathbb{Z}/8$	Cyclic	Yes
	$\mathbb{Z}/4 \times \mathbb{Z}/2$	Dir. prod.	Yes
	$\mathbb{Z}/2 \times \mathbb{Z}/2 \times \mathbb{Z}/2$	Dir. prod.	Yes
	D_4	Dihedral	No
	Q_8	Quaternion	No
9	$\mathbb{Z}/9$	Cyclic	Yes
	$\mathbb{Z}/3 \times \mathbb{Z}/3$	Dir. prod.	Yes
10	$\mathbb{Z}/10$	Cyclic	Yes
	D_5	Dihedral	No
11	$\mathbb{Z}/11$	Cyclic	Yes
12	$\mathbb{Z}/12 \simeq \mathbb{Z}/3 \times \mathbb{Z}/4$	Cyclic	Yes
	$\mathbb{Z}/2 \times \mathbb{Z}/6$	Dir. prod.	Yes
	D_6	Dihedral	No
	A_4	Alternating	No
	Q_{12}	Generalized quaternion	No
13	$\mathbb{Z}/13$	Cyclic	Yes
14	$\mathbb{Z}/14$	Cyclic	Yes
	D_7	Dihedral	No
15	$\mathbb{Z}/15$	Cyclic	Yes

17.2 A_n is Simple for $n \geq 5$

Here, we present a proof that A_n is simple for $n \geq 5$. The proof is both an important part of the classification of finite simple groups, and is readily accessible to us based on our existing toolbox developed in the preceding chapters.

Theorem 17.1. *A_n is simple for $n \geq 5$*

We showed in Chapter (13) that A_5 is simple. We will prove simplicity for $n \geq 5$. Before we begin the proof, we first state and prove two lemmas which we utilize in the proof.

Lemma 1

If $N \triangleleft G$ and K is a subgroup of G, then $K \cap N \triangleleft K$.

Lemma 1 proof: This follows as part of the second isomorphism theorem. To reiterate, from normality of N in G it follows that for any $n \in N$ and for any $g \in G$, $gng^{-1} \in N$. Therefore for any $x \in K \cap N$ and for any $k \in K$, $kxk^{-1} \in N$. Also, since k, x, $k^{-1} \in K$, by closure under group multiplication it follows that $kxk^{-1} \in K$. Therefore kxk^{-1} is in both K and N, and hence is in $K \cap N$. This shows that $K \cap N \triangleleft K$. $\boxed{\text{QED}\checkmark}$

Lemma 2

Any even permutation can be attained by some composition of 3-cycles. In other words, the 3-cycles generate A_n.

Lemma 2 proof: Any element of S_n can be represented as a product of (not necessarily disjoint) transpositions as follows:

$$(abcdefg...) = ...(ag)(af)(ae)(ad)(ac)(ab) \qquad (17.1)$$

For a 3-cycle, this yields:

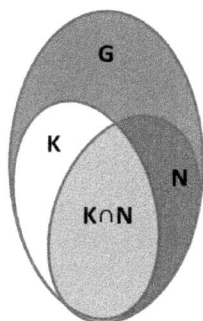

Figure 17.1: Intersection Normalcy: $K \cap N \lhd K$

$$(abc) = (ac)(ab) \tag{17.2}$$

A_n is the group of *even* permutations on n letters, hence any element of A_n consists of an even number of transpositions. We claim that any pair of consecutive transpositions can be represented as a product of 3-cycles. In the case of non-disjoint 2-cycles we have already shown in Equation (17.2) how that can be done. In the case of an arbitrary disjoint 2 cycles, we can represent as a product of 3-cycles as follows:

$$(ab)(ce) = (bce)(bae) \tag{17.3}$$

This concludes the proof that any element of A_n can be represented as a product of 3-cycles. $\boxed{\text{QED}\checkmark}$. For further illustration, two of several possible schemes for representing a 5-cycle and a 7-cycle each as a product of 3-cycles is the following:

$$(bedca) = (bca)(bed) = (eab)(edc) \tag{17.4}$$

$$(bedcagf) = (bgf)(bca)(bed) = (efb)(eag)(edc) \tag{17.5}$$

Figures (17.2) through (17.4) illustrate 3-cycle represen-
tations of some elements of A_n. Figure (17.2) shows a 3-cycle
decomposition of the 5-cycle $(acbde)$. Figure (17.3) shows a
3-cycle decomposition of the 7-cycle $(acfbdge)$. And Figure (17.4) shows a 3-cycle decomposition of a product of
disjoint 2-cycles $(ab)(cd)$. Note that the 3-cycle decompositions are typically not commutative. For example, $(ab)(cd) =$
$(bcd)(bad)$ but $(ab)(cd) \neq (bad)(bcd) = (ad)(bc)$.

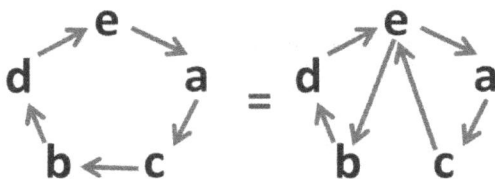

Figure 17.2: A 3-cycle decomposition of a 5-cycle: $(acbde) =$
$(ebd)(eac)$. Note that the decomposition is not unique, e.g.
it is also true that $(acbde) = (ade)(acb)$.

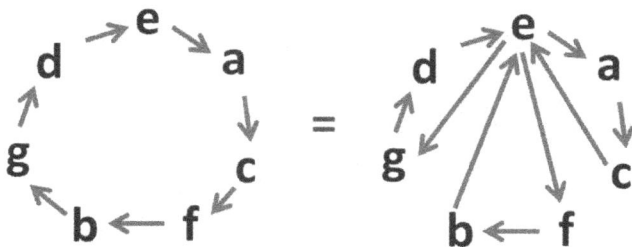

Figure 17.3:   A 3-cycle decomposition of a 7-cycle:
$(acfbgde)$ $=$ $(egd)(efb)(eac)$. The decomposition is not
unique, e.g. it is also true that $(acfbgde) = (cea)(cgd)(cfb)$.

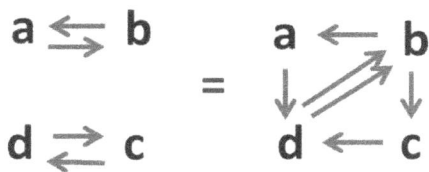

Figure 17.4: A 3-cycle decomposition of a product of disjoint 2-cycles: $(ab)(cd) = (bcd)(bad)$.

Proof of Theorem

Equipped with the above two lemmas and with proof that A_5 is simple, we proceed to prove that A_n is simple for $n \geq 6$. We initialize with $n = 6$, hence A_{n-1} is known to be simple. We will then proceed inductively by showing that if A_{n-1} simple, then so is A_n. Let

$$X = \{1, 2, 3...n\} \text{ and } G \simeq A_n \qquad (17.6)$$

Consider the natural group action $G \cdot X$, defined by

$$\sigma \cdot X = \{\sigma(1), \sigma(2), \sigma(3), ...\sigma(n)\}, \qquad (17.7)$$

where $\sigma \in A_n$. Let us denote the stabilizer under G of the point $i \in X$ by G_i,

$$G_i = \{\sigma | \sigma(i) = i\}. \qquad (17.8)$$

Then for every i,

$$G_i \simeq A_{n-1}, \qquad (17.9)$$

hence by our induction premise, G_i is simple for every i. Now consider some $N \triangleleft A_n$. It follows by Lemma 1 that

$$G_i \cap N \triangleleft G_i \text{ for all } i. \qquad (17.10)$$

Then since G_i is simple, it follows that there are only two possible cases:

$$G_i \cap N = \begin{cases} G_i, & \text{case (i)} \\ \{e\}, & \text{case (ii)} \end{cases} \qquad (17.11)$$

Case (i): $G_i \cap N = G_i$ for some i. i.e $G_i \subset N$ for such i. Then since $G \cdot X$ is a transitive group action and $G_i = Stab_G(i)$, it follows that given any σ such that $\sigma(i) = j$,

$$G_j = Stab_G(j) = \sigma G_i \sigma^{-1} \qquad (17.12)$$

Because $G_i \subset N$ and $N \lhd G$, we have,

$$G_j = \sigma G_i \sigma^{-1} \subset \sigma N \sigma^{-1} = N \qquad (17.13)$$

Therefore,

$$G_j \subset N \ \text{ for all } j \qquad (17.14)$$

Given any arbitrary 3-cycle (abc), we can always pick some $j \notin \{a, b, c\}$. Then for such j, and such arbitrary 3-cycle (abc),

$$(abc) \in G_j \subset N \qquad (17.15)$$

Therefore all 3-cycles are contained in N. And since by lemma 2, the 3-cycles generate A_n, it follows that $A_n = N$. This completes the first part of the proof.

Case (ii): During the proof of case (i) above, we learned that if $G_i \cap N = G_i$ for some i, then $A_n = N$ and therefore $G_i \cap N = G_i \cap A_n = G_i$ for all i. Therefore the statement of Case (ii) can be updated to read:

$$G_i \cap N = \{e\} \ \textit{for all } i \qquad (17.16)$$

This implies that the G-action restricted to N has no associated fixed points. i.e. N fixes no j, hence there does not exists a $\sigma \in N$ and a $j \in X$ such that $\sigma(j) = j$. This implies that if $\sigma \in N$ and $\tau \in N$, then $\sigma(j) \neq \tau(j)$ for any j; else $\sigma\tau^{-1}(j) = j$. Therefore the elements of N do not agree on any letter. This significant constraint should conceivably lead to a contradiction. And it will.

Without loss of generality, let us consider two cases:

1. $\sigma = (abcd)\rho \in N$

2. $\sigma = (ab)(cd)\rho \in N$

Where ρ fixes all letters in $\{a, b, c, d\}$. i.e. $\rho(i) = i$ for any $i \in \{a, b, c, d\}$. Let $\tau \in A_n$ such that $\tau(d) = e$ and $\tau(i) = i$ for any $i \in \{a, b, c\}$. In other words we require, $\tau \in G_a$, $\tau \in G_b$, and $\tau(d) = e$. By our induction premise that $n > 5$, the requirements on τ can always be met. For example, if $X = \{a, b, c, d, e, f\}$, we can pick $\tau = (def)$. Since $N \lhd A_n$,

$$\omega = \tau\sigma\tau^{-1} \in N \qquad (17.17)$$

Evaluating ω for cases 1 and 2 above, we get,

1. $\omega = (abce)\tau\rho\tau^{-1} \in N$

2. $\omega = (ab)(ce)\tau\rho\tau^{-1} \in N$

In both cases we see that $\omega \neq \sigma$ because $\omega(c) = e$, while $\sigma(c) = d$. However, $\omega(a) = \sigma(a) = b$. This is a contradiction. Therefore N contains no non-trivial element, and hence $N = \{e\}$. This concludes the proof that A_n is simple for $n \geq 5$. $\boxed{\text{QED}\checkmark}$.

17.3   Exercises

1. Show that there is only one isomorphism class for groups of order 15

2. Show that $\mathbb{Z}/6 \simeq \mathbb{Z}/2 \times \mathbb{Z}/3$

3. Show that $\mathbb{Z}/mn \simeq \mathbb{Z}/m \times \mathbb{Z}/n$ if $gcd(m,n) = 1$

4. Consider the group $\mathbb{Z}/8$

   a) How many generators has it?

   b) How many elements of order 4 has it?

   c) How many elements of order 2 has it?

5. Show that there are only two isomorphism classes for groups of order 10

6. Show that a group of order 10 is either isomorphic to $\mathbb{Z}/10$ or D_5

7. Classify all groups of order 20

8. Classify all groups of order 30

9. Represent $(ijklmn)$ as a product of transpositions.

10. Represent $(abcd)(ef)$ as a product of 3-cycles.

11. Show that A_n is generated by 3-cycles

12. For $\sigma \in S_n$ show that

$$\sigma(g_1 g_2 g_3 \cdots g_n)\sigma^{-1} = (\sigma(g_1)\sigma(g_2)\sigma(g_3) \cdots \sigma(g_3))$$

13. Show that the quaternion group is a splitting-simple group, i.e. it cannot be expressed as a semi-direct product of its subgroups.

Bibliography

[1] David S Dummit and Richard M Foote. Abstract algebra, (2004).

[2] M Artin. Algebra. featured titles for abstract algebra series, 2011.

[3] Israel N Herstein. *Topics in algebra*. John Wiley & Sons, 2006.

[4] John B Fraleigh. *A first course in abstract algebra*. Pearson Education India, 2003.

[5] Charles C Pinter. *A book of abstract algebra*. Courier Corporation, 2012.

[6] Koichiro Harada and Ronald Solomon. Finite groups having a standard component L of type $\widehat{M}_{12}$ or $\widehat{M}_{22}$. *Journal of Algebra*, 319(2):621–628, 2008.

[7] Arthur Cayley. VII. on the theory of groups, as depending on the symbolic equation $\theta n = 1$. *The London, Edinburgh, and Dublin Philosophical Magazine and Journal of Science*, 7(42):40–47, 1854.

[8] Arthur Cayley. LXV. on the theory of groups, as depending on the symbolic equation $\theta n = 1$.—Part II. *The London, Edinburgh, and Dublin Philosophical Magazine and Journal of Science*, 7(47):408–409, 1854.

[9] Victor J Katz and Annette Imhausen. *The Mathematics of Egypt, Mesopotamia, China, India, and Islam: A Sourcebook.* Princeton University Press, 2007.

[10] Krishnaji Shankara Patwardhan, Somashekhara Amrita Naimpally, and Shyam Lal Singh. Lilavati of bhaskaracarya. a treatise of mathematics of vedic tradition, 2001.

Index

www.ingramcontent.com/pod-product-compliance
Lightning Source LLC
Chambersburg PA
CBHW071650200326
41519CB00012BA/2471